The presentation
of technical information

Reginald O. Kapp, B.Sc.[Eng.], M.I.E.E.
Formerly Pender Professor of Electrical Engineering
University College and Dean of the Faculty
of Engineering in the University of London

The presentation
of technical information

Revised edition
containing an additional chapter
on the presentation of numerical information
in SI units by
Alan Isaacs, Ph.D., B.Sc., D.I.C., A.C.G.I.

Constable London

First published in Great Britain 1948
by Constable and Company Limited
10 Orange Street London WC2H 7EG
Copyright © 1957 Constable & Co Ltd

ISBN 0 09 459070 2

Reprinted 1957
Reprinted 1958
Reprinted 1961
Reprinted 1964
Reprinted 1967
Reprinted 1970
Second edition 1973

Set in Monotype Bembo
Printed in Great Britain by The Anchor Press Ltd,
and bound by Wm Brendon & Son Ltd,
both of Tiptree, Essex

Contents

I **Functional English**   13

The call from industry. The technologist's responsibility towards his mother tongue. Importance of language in all work that requires collaboration. Need for study of functional English from the school to the postgraduate stage.

II **Functional and imaginative literature compared**   22

Styles in literature. Imaginative literature as the language of introspection: functional literature as the language of inspection. The main requirements of functional English. Positive and negative characteristics.

III **The problems of functional English**   28

Importance of maintaining receptivity in the person addressed. Problems of language and style. Problems of logic. Problems of psychology.

IV Choice of material 37

Choice depends on purpose of a discourse. Some problems concerning choice of material. Importance of well defined terms of reference.

V The work done by the person addressed 45

Presentation demands knowledge of mental processes. Need of metaphor in the science of psychology. The mind as a storehouse of memories. The act of associating. The act of understanding. The act of memorising.

VI The proper pace 59

The virtuosity in timing of the music-hall comedian. Importance of pace and timing in the presentation of technical information. Distinction between putting information on record and conveying it from mind to mind. Examples of the wrong pace. Difficulty of conveying numerical information. Importance of being stimulating.

VII What it is about 71

Expanding a simple instruction as an aid to receptivity. Defining the situation. Defining the theme. Reminders and subsidiary reminders of existing knowledge. Defining the method. Forestalling misapprehensions. Transitions. Importance of good planning in a discourse.

VIII Making it easy to understand 93

The imaginative writer implies where the functional writer explains. Comment words. Bridges of logical reasoning. The discipline of facts and the discipline of words. Subsidiary bridges. The presentation of graphs and diagrams.

IX Making it easy to remember 115

Unretentive audiences. Associations as aids to memory. Importance of relevance. Timing. Preparation. Beacon words. Creating a state of anticipation. Judicious repetition. Summaries. Use of analogy. Memorable phrasing.

X Circumlocutions 133

Clumsy ways of defining the theme. Circumlocutions used as substitutes for precise nouns or prepositions. Circumlocutions avoided by recasting a sentence.

XI Generalisations 139

General statements only comprehensible with the help of examples. Confusion between analogies and examples. Importance of choosing examples that are familiar to the person addressed. General statements presented as memorable phrases. General statements presented as theme sentences.

Contents

XII **On meaning what you say** — 146

Importance of a time interval between writing and revising. Careless statement. Overstatement. Wishful statement.

XIII **Qualifications** — 152

Effort required to understand a statement coupled with a qualification. Means of separating the qualification from the statement. Giving warning that a statement is about to be qualified. Replacing a qualified by an unqualified statement.

XIV **Metaphor** — 159

Live and dead metaphors. Error of confusing metaphorical and literal statements. Value of metaphor. Aptness. Freshness. Familiarity.

XV **Words** — 169

Importance of a technical vocabulary. Effect of use of words on habits of thought. Typical examples of carelessness, illiteracy, illogical thinking, ignorance of technical matters.

XVI **Presenting numerical information** — 178

Grammar and style in numerical information. SI units and their use. Conventions adopted with SI units.

Preface

The late Professor Kapp based this book on a course of lectures that he gave in the Faculty of Engineering at University College, London. In the twenty-five years since the book was first published there has been a substantial and steady demand for it. The clear implications are that there was a need for the book and that Professor Kapp's work fulfilled this need.

Recognising that the interest in the book still continues, the publishers suggested that I should revise and update it where necessary. This has been a pleasant and not too difficult task – the book has remained remarkably topical. Here and there changes of emphasis have been required and the new edition seemed to provide an opportune moment to add a chapter on the presentation of numerical information in SI units.

When Professor Kapp originally suggested the course of lectures upon which the book is based, he encountered a good deal of apprehension amongst his colleagues. He was told, for instance, that the whole of the ground ought to have been covered at school. He was also advised to abandon the idea of a special course in presentation and, instead, to use the students' laboratory reports as a basis

Preface

of training in the art of exposition, going over each student's report with him sentence by sentence.

In the preface to the first edition of his book Professor Kapp answered his critics in these words:

'These suggestions revealed a serious underestimation of the magnitude of the problems that the executive engineer or scientist has to solve in the course of his professional work when he is presenting technical information or reasoned argument. These problems simply do not arise in the elementary factual statements contained in a student's laboratory reports. They are too advanced for the school curriculum. They are not problems in grammar and syntax. They are rarely problems in literary style. They are most often problems in logic and psychology. Teach a man to think clearly, and he is likely to express himself clearly; teach him to think about the person addressed, and he will have learnt the first lesson in the art of conveying information effectively from mind to mind. But teach him only how to turn out well constructed sentences and he may fail badly in the art of exposition.

'This is no doubt obvious. But the art of exposition has, for educationalists, not yet become a "subject" and grammar and syntax have. This is why English scholars, even though they be expert in the *practice* of exposition, have not yet given their minds to its *teaching*.

'If they are to do so in future those for whom exposition is of vital concern must give a lead. It is for those, such as scientists and engineers, who have personal experience of the difficulty of conveying technical information from mind to mind, to map out the ground that

Preface

should be covered by the study of the expositor's art. With this in mind I took some pains to clarify my own ideas on the subject and to define and classify those problems in presentation that had given me personally the most trouble.'

These remarks are as pertinent today as they were when they were written. The problem of communication still persists for scientists and technologists. Many of them continue to use an ugly jargon and style in their reports that make their ideas accessible only to those with the patience and determination to unravel their prose. Careful study of this little book will undoubtedly be of great help to all students and exponents of technical subjects.

A.I.

1973

I Functional English

The call from industry

The need for an improvement in the standard of presentation of technical information has long been felt, and has been expressed with some insistence during recent years by a number of learned societies and editors of technical journals. The Institutions of Mechanical, Electrical, and Civil Engineers each call for a wide general education for engineers to enable them to participate in the organisation and management of engineering projects and works.

The idea of an engineer as an applied scientist concerned only with design and construction is no longer acceptable. To follow a successful professional career scientists, engineers, and technologists must be able to communicate with each other as well as with accountants, sales personnel, business men, and politicians, all of whom may be concerned in making decisions relevant to their work.

The pressure for a change in our educational methods comes from men actively engaged in the scientific and engineering worlds. They are in the best position to know what is wanted there, and most of them are, as employers

The presentation of technical information

of university graduates, well acquainted with the existing methods. Such pressure cannot be resisted by those of us who are responsible for the training of the rising generation. We must do what we are asked. We must give more breadth to technical education, even though it mean some sacrifice in depth. We must include, in so far as we are able, those subjects specifically asked for. And among them we must give a high priority to what is generally called English, but what, with a view to defining our terms of reference more precisely, I propose to call the *presentation of technical information*.

The technologist's responsibility

One reason (though there are others) why we should make every effort to improve the standard of presentation of technical information is that, today, the technologist has a big responsibility towards his mother tongue. Never before has it been so great. For nowadays technical literature makes up a large proportion of the nation's output of printed matter. There is enough of it to set the pattern of our national speech. Many of us have to read a score of pages written by scientists or engineers for every one page written by men of letters. Moreover, as a result of the technical and economic leadership of English-speaking nations in the free world, our language has become known in other lands chiefly through our technical literature. To many foreigners normal English will not be the language of Shakespeare, or of the translators of the Bible, or of a Lamb or a Hazlitt or a Shaw or an Osborne; it will be the language in which our

Functional English

scientists and engineers clothe their thoughts. Those of us who are responsible for the training of these can have some influence on the way men will talk in remote corners of the earth.

Our language is a national heritage of which we may justly be proud. We must not allow it to become as drab and uninteresting as our industrial towns. We must do more than merely preserve its grammar and syntax. We must not permit ugly, tortuous, meaningless turns of phrase to become second nature, to be treated as the ideal. We must guard against the operation of a Gresham's Law that would cause such turns of phrase to become common currency. And it is such turns of phrase that often characterise the poor expositor.

It is rather disturbing to think what may happen if we continue to neglect our responsibility. Some of those who have lived in the Far East are concerned that pidgin English may become the foreigner's notion of our mother tongue. But even pidgin English is direct and full of variety in comparison with the language of meaningless stereotyped phrases, involved syntax, cumbrous sentences in which much of our technical literature is cast. It would be less painful to read a translation of Shakespeare into pidgin English than into the technologist's English that would cause Mark Antony to say: 'In this case I have undertaken the journey here for the purpose of the interring of the deceased. From this point of view I do not, however, propose putting anything on record in so far as praise is concerned'.

The presentation of technical information

The technologist as a collaborator

The ugliness of such completely unliterary English is, as I have said already, only one of the reasons why the rising generation ought to be taught to avoid it. Another reason is quite a simple one. Such English fails to achieve its purpose.

This purpose is to influence other people's thoughts and actions, as Mark Antony knew full well. To achieve it thought must be conveyed from mind to mind. This is by no means always an easy thing to do, as will be apparent when we have looked closely at some of the problems involved. Yet civilisation depends on it being done well.

The popular notion of the scientist as a recluse who makes great discoveries in the solitude of his laboratory was, in bygone days, sometimes true of the pure scientist. It is rarely so now. And it has never been true of the engineer. He can accomplish little except in cooperation with others. His day is crowded with talks, conferences, committees. His contacts with people are numerous and varied. In all of them mind addresses mind. It does so, sometimes through the spoken, sometimes through the written word. Talk and paper are, in these days, among the engineer's most important tools. He must learn to handle them well. The executive engineer has a greater use for them than for the tools that are found in the carpenter's and fitter's shops. So why think that these alone are educative? Why train engineers in the use of tools that they may never have to touch again once they have been launched on their professional career and teach them nothing about the tools that they *will* have to use?

Functional English

The bad expositor may, and often does, provide an impressive volume of published work. It may contain a valuable record of profound thinking. But yet it will fail to be very effective. With sublime conceit he thinks himself, perhaps, superior to the obligations of mere craftsmanship; or it has never occurred to him that rather hard work has to be done whenever thought is being transferred from mind to mind; or, if it has occurred to him, he is content to let the reader do the whole of this work, to put into the right order in his mind what is in the wrong order on the paper, to draw the conclusions he is meant to even when they are not stated, to jump without guidance to the significance of a statement, to bridge any gap the author's carelessness may have left in a line of reasoning. The books of such an author are like quarries rich in precious ore but hard to work. 'Let those who want the ore', the author seems to say, 'dig for it'. But will they? Need they?

Sometimes they have no choice. An author with unique and indispensable information has his readers at his mercy. The student who can find no well written textbook must use a badly written one. He must quarry hard in it if he would pass his examinations. Every expert, again, who would know what his contemporaries are doing must spend many weary hours quarrying in atrociously composed contributions to learned societies. For well composed ones are all too rare. And the works manager who relies on his technical experts for guidance has often no choice but to take their reports home with him to read at leisure during long evenings. For he finds it useless to ask for verbal explanations; the spoken words of the experts prove no more illuminating than the written ones. So while he should be recuperating for the next day's task he must quarry instead; quarry among a disordered sequence of ideas, clumsy sentences, unfinished arguments, unexplained conclusions, undefined terms, ambiguous phrases. All of them, the undergraduate, the expert, the works manager, are turned into quarry slaves. The bad expositor is their master.

If he possesses knowledge that cannot be obtained elsewhere he remains their master. For knowledge is, indeed, power. But as soon as a better expositor comes along with the same knowledge the slaves will turn to the new one and be free of their slavery. For in these days, competition for attention is great. Every technical man encounters far more written work than he ever has time to read. He gives his attention only to those who know how to earn it and hold it.

So those scientists and philosophers who neglect the problems of presentation should ponder on the fate of all tyrants in history. If they argue that these problems are too insignificant for their exalted study, if they plead that their time is valuable, they should reflect that the time of the person addressed is valuable too. If he can help it this person will not waste his time quarrying. He will avoid the man who tries to make him do so. In the end the proud scientist or philosopher who cannot be bothered to make his thought accessible has no choice but to retire to the heights in which dwell the Great Misunderstood and the Great Ignored, there to rail in Olympian superiority at the folly of mankind.

Only if an author has little of value to say may he

indulge in an obscure style. Then it may even be an asset to him for it may cause his contributions to learned societies to escape criticism, to be measured by their bulk and not by their content. If an author is a humbug he had better also be unreadable; he may then not be found out.

A subject for advanced study

So much for the reasons why the presentation of technical information ought to find a place in the studies of many scientists and philosophers, and certainly in those of engineers. We must now consider more closely what the nature of this subject is.

It is not an elementary subject. The defects that have called for so much recent insistence on a higher standard of presentation are rarely defects of grammar or syntax. University graduates have reached at least O-level standard in general subjects before they embark on their specialist A-levels. They can fairly be described as educated men and women. If they are bad expositors the reason is not illiteracy. It may not even be lack of appreciation of what is regarded as a good literary style. A person who expresses himself in such a style may yet be a bad expositor. For there are a variety of literary styles, as I shall show later. In some of them obscurity is no defect. Indeed, the beauty that gushes from the spring of a poet's fancy may derive from some enchantment of obscurity. But for the presentation of technical information the least obscurity is a grave defect. So of all the available ones the learned societies have one particular style in mind when they call for a higher standard in

English. It will be convenient if this style has a name. So I will call it *functional English*.

It sounds easy enough. For by functional English I mean the English that any writer uses who expresses his meaning clearly and without ambiguity; who spares his readers unnecessary effort; who selects every item and places every sentence and every word so that it will meet the function assigned to it, just as the designer of a girder selects and places every strut and every tie bar. Functional English presents facts and ideas simply and logically.

If our great masters of prose provide more, no author who wishes to be read and understood ought to provide less. When his sales depend on his readableness he cannot afford to. So even the despised best seller is written in a style suitably described as functional.

The functional style may sound easy, but impressions are deceptive. Though it is not always a very good style it cannot be achieved without hard work. It requires of the writer many qualities, not the least of which are craftsmanship and clear thinking. Without training and practice a person can no more write good functional English than play the violin.

If we are to teach this style properly we shall first have to do a considerable amount of preparatory work. We shall, for instance, have to discover what is wrong with the methods of presentation as now only too often practised by scientists, engineers, and, let me add, a good many philosophers, and we shall have to consider the defects not in terms of vague generalities, but in precise detail. Only so shall we learn by our past mistakes.

And when we have noted the mistakes that are com-

Functional English

monly made it will be our task to classify them and to understand both their nature and their origin. Not till this has been done will it be possible to formulate rules of good presentation and to develop a technique for teaching these rules to others. Many of the problems are, as will become apparent later, too advanced for the school curriculum. Many of them can only be properly appreciated by the mature mind of the postgraduate.

II Functional and imaginative literature compared

It would be a mistake to suppose that, in order to train a student in the use of functional English, it suffices to make him acquainted with the best in English literature. For most that is best in English literature is not written in the style that I am calling functional. There is, it must be remembered, no limit to the number of possible styles in literature, just as there is no limit to the number of possible styles in architecture. Each is adapted to its setting and purpose. Functional English, like functional architecture, is but one of the many possibilities.

The architecture of a mediaeval castle, with its crowded turrets irregularly placed, looks right in mountainous country. The many rounded surfaces seem to get in each other's way just as the hills do. In flat open country a style that employs a spacious symmetry with long clear horizontal lines seems more appropriate. Rich ornament, again, is suitable for a palace, one purpose of which is to provide a background for the pageantry of ceremonial occasions. It would look silly on a factory or an office building. So one would not advise an architect of factories or office buildings to devote all his time to the study of mediaeval castles and renaissance palaces.

In literature there is one style for lyrical poetry, another

Functional and imaginative literature compared

for epics, another for descriptive prose passages, another for oratory. The style that is apt in humorous writing is very different from that employed by the serious essayist. There is even a subtle distinction between the conversational styles most appropriate to the novel and the stage play. Each of these and the many other styles to be found in literature presents its own problems and can only be mastered after careful study and some practice.

Functional English is the proper vehicle for a report to a superior or an instruction to a subordinate, for a paper to a learned society, for a lecture or a book on any technical subject. It has its uses in the committee room and the office. It is the language in which much good history is written. It is to be found occasionally in good fiction. The best adventure stories are, for instance, partly written in functional English. *Robinson Crusoe* is a classical example. It is the proper language whenever bare factual information has to be conveyed either by speech or by writing. And it is also the proper language when inferences are being drawn from the facts; when the results of a scientist's subtle and penetrating reasoning are being communicated. It is the language of logic, of all argument in which the appeal is to the intellect and not, as in poetry, to the imagination, or, as in oratory, to the emotions. It is the language that philosophers ought to use, but alas, rarely do.

The purpose of functional English can be stated in a few words. It is always to convey *new* information. The information may consist of all kinds of facts, of inferences, arguments, ideas, lines of reasoning. Their essential feature is that they are new to the person addressed.

The presentation of technical information

This is obvious. For functional English is the language of collaboration. Those collaborating must tell each other what to do. They must give reasons and explanations. Sometimes they must debate and argue. When all this happens the important things said are, of course, mostly those things that the persons addressed did not know before, or had, at least, overlooked or forgotten. For it would be a waste of time and breath to tell people only what they knew perfectly well already. But is it not always the function of language to convey new information? Can one justify the use of language if it does not?

One can. And it is important that this be realised. For so long as functional English is regarded as synonymous with any good style its scope and purpose will be missed and its peculiar problems will not be faced or overcome. When a poet tells us that the spring is a beautiful season, that the birds please with their song, and that lovely flowers bloom in May, he conveys something for which neither the words 'new' or 'information' are appropriate. We knew it all before we were told. Yet the poet does convey something. Perhaps the word 'appreciation' would serve to define it. The poet reminds us of what we have already experienced. With the artful use of words he enables us to experience it again and more fully than ever.

So it is with much of our best literature. Lyrical poetry is not the only example. Art, as someone has said, is emotion recollected in tranquillity. And hence what is finest both in prose and in poetry rarely attempts to convey much unknown information and but rarely calls for the use of functional English. We accept what our great

Functional and imaginative literature compared

masters of writing tell us, not for its novelty, but because we feel that it is true, because we feel that we have always known it, though it may be but dimly. What the great poet or prose writer does is to make our dim knowledge fully, vividly conscious.

It would not do so if it conveyed *new* information. To a person who had never experienced the spring a poem about its beauties would mean far less than it does to a person who has had the experience. It is significant that on reading fine poetry one is not at all inclined to exclaim 'how new this is'. One is more likely to say, as one does of what one knows already, 'how true this is'.

This contrast calls for a label by which all that mass of literature can be collectively named that does not come under the category 'functional'. I suggest 'imaginative literature'. This like any such label must necessarily be a little vague and inaccurate, for the whole range of literature that is not functional is far too great to be compressible into one single descriptive word. Nevertheless 'imaginative literature' may serve the purpose of the moment.

There is at least one feature by which one can distinguish functional from other literature fairly sharply. It is rather a profound feature and deserves a moment's attention if functional English is to be appreciated at its proper value, neither too high nor too low. Imaginative literature is largely concerned with ourselves, with our thoughts, our feelings, our reactions to experience. It is the literature of insight. When it discusses the outer world it is not so much in terms of what things are in themselves as of what they are to us. By contrast functional English is chiefly the language of science. It is nearly always concerned with the outer world; it rarely conveys information obtained by introspection.

The English used in imaginative literature, the English that awakens self knowledge, that conveys the result of introspection, the English of insight, employs word music, imagery, metaphor, symbolism. A skilful writer makes cunning use of understatement, overstatement, oblique statement, play on words, startling juxtaposition. He introduces into his descriptive passages sundry significant small details with the sole purpose of stimulating the visual imagination. By the use of evocative allusion he creates a desired mood; he plays on the reader's feelings. He masters a hundred subtle means to pile into a few sentences far more meaning than they might be expected to bear. He sometimes refrains on purpose from being explicit. Instead of drawing inferences he leaves the reader to do so. By all kinds of artful device he arouses the expectancy of one thing, only to surprise, dramatically, with something else. Much good imaginative literature may be exuberant, colourful, precious, sonorous, ornate. Certainly not as simple, clear, and concise as functional literature has to be. The writer of imaginative literature is a magician, who confers on words vast powers, so that they cease to be mere vehicles for information and become for the time being the masters of men's souls.

How different is the use of words in functional English. A style that is magnificent when the words tell what any man can discover who looks deeply enough into himself, is quite wrong when the words tell what can only be

Functional and imaginative literature compared

discovered by observation of the outer world. Then word music only distracts. Imagery confuses. Instead of being used with profusion metaphor must be used sparingly; it may prove misleading. Understatement, overstatement, oblique statement are all out of place; they are all the same thing as mis-statement. Plain statement alone serves the purpose of functional English. Anything that might create a mood would only spoil the reader's receptivity to facts and logic. Words must not be made to carry either more or less meaning than they do in common usage. If any expectancy is aroused it is usually a mistake to disappoint it; to do so deflects attention.

Yet functional English is not achieved merely by leaving out everything that makes for the glory of imaginative literature. It has positive as well as negative characteristics. The writer of functional English is a magician, too, though a minor one. If not masters of men's souls, his words are, at least, masters of their minds. As he writes he confers on the words the power to make those who read him think as he wills it.

III The problems of functional English

The aim of a good functional style is to maintain receptivity in the person addressed.

To maintain, I say advisedly; not to create. For receptivity cannot be stored like the energy in the spring of a clockwork engine. One can wind up the engine and place it on its rails. It will then run for quite a while without further attention. But one initial stimulation of the reader's interest does not suffice to maintain receptivity for page after page of difficult subject matter. This can only be done by a continuously renewed effort and success depends largely on the writer. A bad expositor may, by his bungling, do much to cause the reader to lose receptivity, while an expositor who has successfully solved the problems of functional English will know how to ensure its maintenance, even when circumstances are not favourable.

These problems are all, by their nature, concerned with the mind of the person addressed. They are all psychological in this sense. But in order to maintain receptivity in the person addressed an author has to solve a number of problems of which the psychological purport is not immediately apparent. It will, in particular, be convenient to distinguish between problems in linguistics,

The problems of functional English

problems in logic, and problems specifically in psychology.

Linguistic problems

The need to be simple, clear, and concise is often emphasised. So much indeed that one might sometimes think that simplicity, clarity, and conciseness were the only requirements of a good functional style. How they may be achieved is a problem in linguistics, and a stiffer one than one might perhaps think.

There are others. Much can be done to lighten the reader's burden if attention is given to such matters as choice of words, grammar, syntax, punctuation, paragraphing. A reader's receptivity is impaired if words are used that are unfamiliar, or vague, or uninformative, or ambiguous. So it is if he has to unravel a tortuous syntax. Misleading punctuation is also harmful. It introduces uncertainty as to an author's meanings: any oddities in punctuation divert the reader's attention from the subject matter. When a change of subject fails to coincide with a change of paragraph the reader becomes confused.

A good expositor must solve yet other problems in linguistics, even though many scientists and engineers seem to think them beneath their notice. They are problems in style. For any lapse from good literary taste is unfunctional. When presenting technical information an author must avoid clumsy phrasing, cacophonous syllables, ugly rhythms. He must not let the reader be jolted by sentences that end too abruptly. He must not let the reader lose the thread of the argument in sentences that drag on for too long. He must take pains to eliminate from his writing anything that might interrupt the smooth flow of thought that he should induce in the reader.

If the person addressed is a scientist or an engineer he may, admittedly, not know the difference between a smooth elegant style and a clumsy one. But one must not think that, because he is too insensitive to notice such things, they do not matter. Even those lapses that pass unobserved do some harm. The person addressed is affected by them, whether he notices them or not, much as he is affected by the flies that buzz around his head. When he is too intent on what he is doing to be aware of the troublesome flies they still disturb him, and impair his efficiency. And this is what a style that is clumsy or in any way insistent does to the reader of technical literature. An author has achieved the perfect style when there is nothing to show the care that he has taken with it. He cannot reach this ideal without solving many problems in linguistics. To do so requires imagination and constant vigilance.

Problems of logic

The first logical *don't* that I would put to those who have technical information or reasoned argument to present is this. Don't allow anything that is irrelevant to intrude into your discourse. It seems an easy *don't* but it is really a very hard one. Testing for relevance raises some of the most difficult problems that an expositor ever has to solve. It is important to remember with all humility that everyone of us fails to solve them very

The problems of functional English

often. The best we can do, is to be aware of our weakness, to be every watchful, and to hope that in time we may improve.

I am surprised that the importance of being relevant is not more often pointed out in dissertations on scientific method. Testing for relevance calls for all the logic that the clearest thinker has at his disposal. Its importance goes far beyond the needs of good exposition. However, even in this narrow field it is essential. If the expositor does not test his statements for their relevance at every turn, the person addressed may well miss what *is* relevant. There are few things more destructive of receptivity than to have to pause during one's reading and ask oneself: 'What has this to do with it?'

Other problems in logic are revealed when one considers a number of the more obvious *don'ts* that should be observed by every expositor. Don't present without adequate explanation ideas with which the reader is unfamiliar. When their presentation is unavoidable their logical connection should be shown with things with which he *is* familiar. Don't omit essential steps in your reasoning. Don't elaborate the obvious. Don't communicate a significant fact or figure without communicating what the significance is. Don't arrange facts and ideas in an illogical order; it places an unnecessary strain on the understanding. Don't allow the emphasis of your discourse to fall on facts of secondary importance; if you do, matters of primary importance may be overlooked and the person addressed will draw the wrong conclusions. Such *don'ts* can only be properly observed if a number of problems in logic have been solved. So they

The presentation of technical information

will have to occupy our more detailed attention in later chapters.

These are some of the negative aspects of the logic of presentation. There are also positive ones. Logic serves the person addressed as a map serves the traveller. The map helps him to know from moment to moment where he is, how he got there, and in which direction his path lies. It enables him to distinguish between main and secondary roads, to discover alternative routes to the same objective, to learn where a stream may be forded and how to avoid those apparent short cuts that lead to a quagmire.

If an author who is presenting technical information does not provide the equivalent of such a map, the person addressed must do so himself or lose his way. He must rearrange in his own mind the material presented to him; he must build any bridges of reasoning that the author has failed to provide; he must occasionally sit back and ponder 'what does this signify?'; he must turn back the pages in search of some half forgotten statement. In short, he must do some of the author's work. How to provide a serviceable map, how to reveal the logical frame on which a discourse is constructed, that constitutes one of the logical problems with which the expositor is faced.

Here again the problems of functional English differ from those of imaginative literature, in degree, if not in kind. In the latter, logic may or may not be important for the *subject matter*. It is rarely very important for the *technique of presentation*. Many famous literary men have proved to be quite illogical in their ways of thinking and yet they have written beautiful prose. They have

The problems of functional English

given great poetry to the world, they have shown profound insight as novelists or dramatists, they have commented as critics revealingly on literature and the arts. And with all their gifts they would, I fear, have failed on the lower level of functional English. I doubt if a great master of English, lacking a sense of logic, could write a good elementary textbook on physics, however well briefed he might be for the task.

Psychological problems

The problems of functional English that are specifically psychological are concerned with the mind of the person addressed. So the first task in solving them is to become acutely aware of that person. In order to emphasise his importance I propose from now on to personify him. In future he shall be called John Smith. To give him a name will also avoid that cumbersome phrase 'the person addressed'.

John Smith may be the reader of a report, a paper on the latest scientific discovery, a newspaper article, a book. He may be a member of the audience at a popular lecture or one of a small band of experts at a meeting of some learned society. He may be a managing director who is receiving an account of the day's happenings from one of the heads of departments or he may be a workman who is receiving his instructions. He may be an undergraduate in the classroom or a learned counsel whom a scientific expert is briefing. His real name, for all I know, may be Mary Smith. For he may be anyone, young or old, male or female, wise or foolish, learned or uneducated, to whom new information or reasoned argument is being presented.

I propose in future also to avoid reference to the writer, the speaker, the lecturer, the author. Whoever he may be, let him be called 'you'. 'You' may be a scientist, a philosopher, a teacher, a journalist; you may be lecturing to a class of undergraduates or to an audience of experts; you may be addressing a superior or a subordinate or a colleague on your own level; your discourse may be written or spoken. Provided only that you are entrusted with the task of communicating new information to John Smith you will, in these pages, be called plain 'you'.

The nature of those of your problems that are specifically psychological depends partly on your subject matter but chiefly on John Smith. The more you know about him the better you will fulfil your task as expositor, and if you ignore him altogether you will not deserve the title expositor; you will be a mere soliloquiser.

It is, therefore, your duty, whenever you are presenting technical information, to keep John Smith well in the forefront of your mind. You must consider carefully the extent of his knowledge, his range of interests, his likes and dislikes, his capacity for understanding, his limitations, the rate at which his mind works, any misapprehensions that he may entertain, any prejudices from which he may suffer, any peculiarities, whatever they may be, that might influence his receptivity for the information that you have to impart.

It is difficult to take so much into consideration, you may object, particularly if there are thousands of John Smiths into whose hands the fruits of your brain are to be

The problems of functional English

placed and if you have never met one of them. I agree. The expositor's art is difficult. It is my concern to make it clear that there is a great deal in it, not to mislead you into the belief that it is easy and can be attempted without trouble or preparation.

The psychological problems are of varying difficulty. The easiest occur when John Smith is a willing co-operator. He usually is. Whether he be a layman who would like to understand scientific matters, a student new to his subject, an expert in search of the latest discoveries in his own field of study, a chief executive who seeks the advice of his staff, he will do his best to maintain receptivity throughout your discourse. If he loses it the fault is probably yours. Somewhere you have bungled. You may have forgotten that it takes two to maintain receptivity.

Greater difficulties arise when John Smith fails to do his part of the work, when he maintains his receptivity grudgingly. The most difficult occasions of all occur when he opposes your effort by a counter-effort. It may happen if what you have to tell him is unwelcome, as it may be for a variety of reasons. Perhaps it is in conflict with his own best interests; perhaps it proves that something that he has said is wrong; perhaps it is emotionally disturbing; perhaps it has humiliating or painful associations; perhaps it refutes some belief that he holds passionately. For even if John Smith is a scientist he may hold some beliefs passionately, unwilling though he is to admit it. Human nature being what it is, John Smith will then, far from making an effort to understand what you have to tell him, make an effort to misunderstand it.

The presentation of technical information

For he has acquired a vested interest in error. Scientific and philosophical controversy abounds in examples of unconscious, but none the less deliberate, misunderstanding: both religious and political controversy provide even more examples.

These are the occasions that test the expositor's art to the full. They call for all the insight he can muster; for all his tact, subtlety, skill, persuasiveness, ingenuity, not to say downright cunning. I have said that it takes two to maintain receptivity. Well, sometimes you have to do the work of two. A skilful counsel does so when he is addressing a jury prejudiced against his case. An orator does so when he persuades an audience to replace preconceived notions by new ones. This is what Mark Antony did in his famous speech to the Roman mob.

Every one of us has had the experience, when in the position of John Smith and in receipt of unwelcome new information, that we rejected it on first hearing. By a counter-effort we kept our receptivity in abeyance. Have we not realised afterwards that we should not have made the counter-effort if the information had been presented differently? So we ought to be ready to appreciate that the devices that belong to the stock-in-trade of barristers and orators are also necessary tools of every expositor.

IV Choice of material

I do not know whether examination papers still contain any questions of the type: 'Write down all you know about Queen Elizabeth'. Examiners who set such questions either overrate the number of words that the candidate can write down in the time allowed or they underrate the amount of knowledge that he ought to possess. A candidate who knows no more of a given subject than he can write down in an examination paper ought to fail.

If John Smith is your chief and he asks you for a report he probably wants it to help him in reaching a decision on a specific point; he does not regard it as an opportunity for you to parade your learning. So do not set out to write down all you know on the subject. You must not provide more than will serve your chief's purpose.

Nor must you, of course, provide less. You must use your judgment as to how much is required. The task of preparing even a short and simple report for your chief may confront you with the problem of choice of material. So may the task of giving an instruction to a subordinate. If he is to do what you want him to properly you must tell him certain things that he ought to know and yet must not confuse him with unnecessary facts. The same problem exists usually in a more difficult form for anyone who sets out to write a paper or a book. It is important to realise that the problem is real and that it rarely solves itself. Often it calls for much care and anxious thought. So a few hints on choice of material may be helpful.

My first hint is this. *Do not begin to select your material until you have found answers to these two questions. What is the information for? Whom is it for?*

Selection of material for a report or a paper

The purpose of the information may be either to influence someone's action or someone's thoughts. An instruction belongs clearly to the former class, a textbook to the latter. Most reports also belong to the former class. They are the basis of immediate decisions. So it is a mistake to include in them anything that cannot influence the decision.

A few other mistakes may be listed as they are not uncommon. It is a mistake to give an unnecessary amount of detail. The result is only confusing. It is also a mistake to give precise figures when round numbers serve equally well. It is a mistake to give weak reasons for a conclusion that has already been supported by one or two strong ones; it tends to leave John Smith with the impression that none of them amounts to much.

And it is often, though not always, a mistake to include alternative, but rejected, solutions of a given problem. If the report is to be the basis of immediate action, John Smith, who looks to the report to help him in a decision, is not much interested in what is *not* to be done; he only wants to know what *is* to be done. Here a meticulous

Choice of material

scientist may err through over-conscientiousness. Having learnt in the school of scientific method that no conclusion ought to be accepted until every alternative has been considered, tested, and rejected he inclines to put precept into practice on every occasion, even out of season. Rejected arguments and conclusions should not be among the material selected in a report to your chief merely as offerings to that exacting god, scientific method, nor merely in order to prove that you have done your job thoroughly. It is wiser to assume tacitly that John Smith, your chief, takes this for granted. But rejected arguments and conclusions must be included if their omission could mislead John Smith or if it could raise doubts in his mind.

These considerations lead me to an allied problem of selection that sometimes arises when an account is being given of some piece of original research work. A number of circumstances have, let us suppose, called for much care and anxious thought, together with not a little trouble and wasted effort. Among them there may have been a search for the right type of apparatus, difficulties in the calibration of the apparatus, problems in the preparation of specimens, disturbing secondary effects, difficulties in the elimination of errors, as well as other apparently trivial matters that could not be ignored. Few such troubles and difficulties are very interesting to the John Smiths who will read your account of the research. They are only anxious to know the results. Yet even when it cannot possibly help other experimenters and can only bore them, the inclusion of such material *may* be justified; for the question may arise at some future date as to whether your experimental methods were sound. Someone may doubt the general applicability of your discoveries. He may say that you only found out something about the particular apparatus that you employed, or that your results are no more than a function of the methods used. Such unkind things are sometimes said, and they are sometimes true.

Now if full details are given of the methods and the apparatus, even apparently trivial details, these questions can later be disposed of. So a wise research worker will sometimes include a considerable amount of tedious detail. But when preparing the results of his work for publication he is often in a dilemma on this account. Shall he spoil the readability of his book or paper by including material that serves the sole purpose of meeting possible criticism? Or shall he select for publication only what is of undoubted interest, with the risk that at some future date, if his results are called in question, there will be no means of allaying suspicions? What makes his decision the more difficult is that he cannot know whether the suspicions will ever arise; he may spoil the readability of his publication for nothing. So the decision must be made for every occasion on its merits and calls for a nice prophetic judgment. All that I can add concerning this dilemma is that the decision will be sounder if the research worker faces his dilemma squarely, and gives his whole mind to it, than if he gives it little or no thought.

Selection of material for a book

When you are presenting technical information in a book, the problem of selection is a little different from when

Choice of material

you are preparing a report or giving an instruction. The purpose of a book (and of a paper to a learned society) is rarely to influence someone's action, at least his immediate action. It is definitely to influence people's thoughts. So the author of a book or a paper to a learned society has a wider range of choice of material. Any knowledge, provided it be true knowledge, influences men's thoughts in the right direction. From this it might be argued that for the preparation of a book or a paper no problem of choice of material arises.

This conclusion would, however, not be correct. Bare knowledge may, admittedly, have some good influence on men's thoughts; even the knowledge to be gleaned from the telephone directory and the railway timetable. The knowledge to be gleaned from an encyclopedia or tables of scientific data and formulae certainly serves in this way. But I am assuming, when giving these hints, that you have undertaken the task of providing something more than *bare* knowledge, that you have set out to provide *assimilated* knowledge.

Those who might be expected to do this do not, alas, always attempt it. A scientist has, let us say, devoted the work of some years to an important branch of his subject. He decides to publish in a book what he has learnt about it. This is, without doubt, worth publishing. But the author is a busy man. He must, by force of circumstances, relegate all literary work to his evening hours when he is tired. Then he succumbs all too readily to temptation and takes the line of least resistance. With a minimum of connecting script he strings his notes together. I will give him credit for placing them in a rough logical order, but

The presentation of technical information

not for much more. The result is certainly not what a book should be; it is more like a ragbag containing items of bare knowledge. The fact that the items are valuable does not suffice to redeem it. Instead of trying to solve his problems of selection the author ignores them. Asked what the material presented was for, he could have given no more precise answer than: 'For publication'.

You must define the purpose of the presentation of any technical information more closely than that, if it is to influence people's thoughts effectively. The expression commonly used to define such purpose is 'terms of reference'. So a rule for the choice of material can be: *closely define your terms of reference before you begin to choose your material. Take good care to include anything that comes within the framework set by these terms and reject anything that does not come within the framework.*

The rule imposes some discipline. For some of the things that come within the framework and should be included may not be readily available. In order to provide them you may have to undertake considerably more preliminary research than you had anticipated. Again, the rule will often require you ruthlessly to reject much valuable and interesting material for no other reason than that it does not come within the framework.

If this rule, simple though it sounds, is not often followed one need not be surprised. It is much to expect a scientist to submit to its discipline. It may cause him to delay publication of valuable material by weeks or months, even by years. It ties him to his writing desk for more hours than he cares to spend there, and keeps him away from his laboratory, where he is longing to be. It

Choice of material

is only too true that many a scientist really does have something better to do than to labour painstakingly to ensure that every one of his publications falls within well defined terms of reference. Proper standards of presentation, in this as well as in other respects, may be practically unobtainable in the scientific world. But if so, do not let us accept complacently what we have to put up with. Do not let us come to regard unavoidably low standards as perfect. Let us learn to discriminate between high and low standards, even though the high ones may be rarely achieved.

As an example of the need for terms of reference let me invent one from the field of history. Its bearing on the problem will then be more widely appreciated than would an example chosen from some specialised branch of science. Suppose that some historian has made a special study of the Treaty of So-and-So. He decides to write a book with that title. Should it contain all that he knows about the Treaty of So-and-So? I am sure not. If it did it would bring knowledge without a context, bare knowledge, of comparatively little service. At best the book would contain the raw material from which a better historian could select what he might need for a more valuable book. In this one would find assimilated knowledge and well defined terms of reference.

There is no limit to the possibilities for such terms. A book might be written on the way the Treaty of So-and-So led to the War of Such-and-Such, or how it revealed the interplay of secular and religious forces, or in what respect the Treaty was an expression of the spirit of the age, or what social changes resulted from it. But whatever

The presentation of technical information

the theme, the title need not be anything but 'The Treaty of So-and-So'. What is important is that the terms of reference must be defined in the author's mind in far more detail than can appear in a title. What is enough for guiding the reader is not nearly enough for disciplining the author.

The same holds if the subject is scientific. The purpose of writing may be only to fill a gap in existing literature on a given subject. If so the author should carefully survey the extent of that gap before selecting his material and keep the ground to be covered well to the forefront of his mind all the while. And if his self-imposed terms of reference are deeper and more specific it is even more important to keep strictly to them.

There is no limit to the possible terms of reference in science any more than there is in history. A scientist may write to prove a given theory, to establish a connection between hitherto separate fields of study, to develop a new point of view, to provide a textbook at a given well defined standard, to enlighten the lay public. All these and innumerable other purposes raise their problems in the choice of material and the problems range in difficulty. The easiest ones occur when the purpose is to give a straightforward account of a rounded-off investigation, the hardest when new ground is being broken in some little explored field of study.

V The work done by the person addressed

If you are to convey technical information effectively to any John Smith you must know what work he is doing while he is receiving it and how to lighten that work for him. This is obvious. But it is one thing to know that it has to be done and another to know how to do it. It would be easier to devise rules of presentation if some expert psychologist had already analysed and classified the work done by a person who is reading or listening to new technical information. But if it has been done at all it has not, so far as I have been able to discover, been done in a form that would serve as a useful guide to the expositor.

Until experts tell us more of practical value about this aspect of mental processes those who have new facts and arguments to present must do their best with such help as their common sense may provide. This tells of at least three acts performed by a John Smith while he is receiving information. They are, respectively, the act of associating, the act of understanding, and the act of memorising. A good expositor can make all three easy; with a bad expositor they all become very arduous. It is, therefore, worth the while of anyone who would learn how to present technical information to give some detailed consideration to the processes of associating, understanding, and memorising. But before I say anything about them I must digress for a moment and say a few words about the terms in which these mental processes are to be discussed.

The metaphor of a storehouse for memories

At the outset I am faced with a dilemma. It faces everyone who has to discuss psychological phenomena. This is due to the lack of a suitable language. In consequence, psychologists inevitably find themselves driven at every turn to the use of metaphor. One need only glance at any book on psychology written by a leading authority to discover the amount of metaphor it contains. It would seem that without the use of this device nothing much can be said about mental processes. This holds for all schools of psychology.

I think that there is a rather profound reason for this. Language was not evolved by the human race in order to discuss what goes on in men's minds. It was designed as an instrument of the cooperative practical work of every day. In the early ages of the human race men talked to each other about what they saw and heard, about what they did, and what they wanted others to do; hardly ever about what they thought or felt deep within their souls. And today language is still mostly used for similar purposes. It serves to communicate the result of inspection, not the result of introspection. One need not be surprised, therefore, that language should prove but a poor instrument for the latter purpose. But it is not such a poor

The work done by the person addressed

instrument when helped by metaphor as when not so helped. If every psychologist has found this out in recent times, every poet found it out at the dawn of civilisation.

So I propose now, without apology, to introduce the metaphor of the mind as a storehouse for memories. To make the metaphor complete I shall also introduce the rather fantastic notion of a storekeeper who works there. I know full well that this, like all metaphors, is inexact. It cannot fail at times to suggest some things that are not quite true. But it will, I hope, convey some things that are true, better than any other linguistic device could do.

The act of associating

If new information is to mean anything at all to John Smith, he must bring it into association with things that he knows already, with things that he has, metaphorically speaking, placed in the storehouse of his memories. This is why the first of the acts that he must perform is the act of associating. During this act mental work is being done and I shall try to give a concrete picture of this work by saying that the storekeeper goes to the shelves on which memories are stored, takes them down, and brings them to a place in the conscious compartment of John Smith's mind, where they are then on display ready for use.

Let a few examples illustrate why associating is necessary. They do not have to be examples in which the information conveyed is scientific. In fact, examples in which it is not so are, I think, more telling. The storekeeper is called upon to perform his various duties when *any* information is being received, no matter whether it be technical or not.

Suppose that John Smith is reading a story about sailing a boat. He must call to mind some of the things he knows already about boats, about nautical terms and about the art of sailing. If he has never learnt anything about such matters, or if he can no longer recall what he has learnt, he will not be able to understand what he is reading. He will probably not remember a word of it. So it is clear that the acts of associating, understanding, and memorising all depend on the storekeeper's activities.

Similarly, when Mary Smith is being told how to knit a new kind of jumper she must recall the difference between purl and plain, together with numerous other things about knitting. When John Smith, again, hears mention of saddling a horse he must call to mind what a horse looks like. When he is studying a treatise on differential equations he must recall his knowledge of the calculus. If he has no knowledge of it the words and formulae placed before him will be meaningless.

Suppose, to take yet another example, that you are explaining a new system of bidding in bridge. John Smith must recall the rules of the game, the method of scoring, the value of a trick in each suit. He must also recall some simple properties of the number 13, for instance, that it is three fours and one over. For many of the subtleties of bridge depend on the various ways in which the number 13 can be reached. He must also recall bidding systems with which he is familiar in order to compare them with the new one. He must recall hands that he has played in the past and consider whether the

The work done by the person addressed

new bidding system would have served him better. Perhaps he must recall his partner's comments on a previous occasion. And he must, at the same time, recall things that you have been telling him and that have been placed in the storehouse five minutes ago. For he will, from time to time, have to be reminded of things that he has just learnt. Were one to make a list of all the distinct items that the storekeeper must bring together and display for John Smith's conscious consideration while you are explaining something quite simple to him, one would have to enumerate literally hundreds of items.

The items are not all on adjacent shelves, and the connection between them would be far from obvious to anyone who did not know the purpose for which they were being assembled. Could he watch the storekeeper at work he might well think him quite irresponsible. Suppose that on some occasion this metaphorical servant takes down, unwraps, and assembles together in the conscious compartment of John Smith's mind the following items: the recollection of an old master seen twenty years ago in the National Gallery, a chemical formula noticed once in a scientific journal, something learnt as an undergraduate about the refraction of light, and a remark overheard yesterday concerning the manufacture of shellac. Would the storekeeper be acting methodically or like an undisciplined monkey?

It would depend entirely on what was engaging John Smith's mind at the moment. If he were reading a treatise on preservative varnishes for oil paintings the storekeeper would be doing his work well. It would

The presentation of technical information

then be his duty to bring to John Smith's consciousness the recollection of the state of preservation of some old master noticed long ago, and to associate it with the chemistry of varnishes prepared from shellac and their light-refracting properties.

The long shelves of the memory have, in the course of years, accumulated vast stores. Whenever we are thinking about anything, no matter what it be, items are being drawn from widely separated parts of these shelves and assembled together. They are of every conceivable variety, ranging from simple facts of everyday life to recondite discoveries in advanced scientific subjects, from those placed on the shelves during early infancy to those placed there but a moment ago. The items include things heard, seen, and felt; tastes and smells; spoken and written words; trivial things and momentous ones; things experienced in reality and things experienced in thought only; observations, sensations, emotions, impressions, opinions, ideas, dreams.

To produce these as and when required, the storekeeper must travel backward and forward over the span of a lifetime, hither and thither from one class of subject to another. How is human memory and the process of recall related to man-made data storage banks and computer systems?

The answer is that at present we do not know, but it seems unlikely that they even work on similar principles. Metaphors must not be driven to the point of absurdity. Though they are helpful, even necessary, in order to describe *what* happens, they are wholly inadequate for the task of discovering *how* it happens. It may be that

The work done by the person addressed

man can never hope to understand in all its aspects the process of recalling.

A preconscious process

What makes it well nigh inaccessible to scientific study is that most of it is unconscious. It is, no doubt, not as fully unconscious as the processes by which food is digested or cells are renewed. It is not as fully unconscious as the processes of the mind that are studied by the Freudian school. Probably it is usually only what is technically called preconscious. It is certainly not near enough to consciousness to be directly observable by introspection and it is obviously not observable by inspection. So we have full knowledge of the *result* only of recalling.

And we take the result for granted. It does not occur to most of us to ask any questions about it. That, whatever the occasion, all the varied associations needed should appear in consciousness just when they are required, is as familiar to us as the act of breathing. And what is familiar never seems to puzzle the non-scientist.

So it was at one time with the falling of apples. The phenomenon was so familiar that no one thought to ask questions about it. Not until that afternoon when Newton lay in his orchard, if the popular story is quite correct, did anyone realise that the falling of apples was a mystery and needed to be explained. Perhaps some psychologist will do one day for the mystery of mental associating what Newton did for the mystery of gravitation.

The presentation of technical information

Until that happens we shall have to be content with the bare fact that the act of associating is not accomplished without the expenditure of effort and time. There are occasions when we are fully conscious of the effort. Then we know that something packed away long ago in the storehouse of our memory is important and that we shall need it for consideration of the subject that is at the moment claiming our attention. But we have to think hard and long before we can manage to recall it.

It is not always so. Most acts of associating are performed so easily and quickly that the effort expended on them is no more conscious than is the work of the storekeeper himself. It is probable, however, that even then an effort is made. Only the effort is below what psychologists call, with one of their many metaphors, the threshold value. It would seem that this threshold has to be exceeded before the effort of recalling something passes out of the preconscious into the conscious compartment of the mind.

If the threshold value is not exceeded the time taken for acts of associating is so short that the work seems to be done instantaneously. The storekeeper does his work 'quick as thought', he is quicker than Puck, who put a girdle round the earth in forty minutes. But even thought must be allowed some time to develop and the speed of thought is, like the amount of work expended on thinking, a variable quantity.

The amount of time and effort depends partly on the person receiving the information. When he is tired he recalls things only slowly and with difficulty. If, moreover, the associations that have to be produced are

The work done by the person addressed

painful or humiliating they are not brought forward easily. In another of their metaphors, psychologists say that such associations are opposed by a resistance. This is what I referred to in Chapter III as a counter-effort.

The amount of time and effort depends also largely on the subject matter. The ease with which any particular item can be recalled from the storehouse for memories is among the subjects that have already received the attention of psychologists. Apart from the resistance to recalling mentioned above, there are many other circumstances on which memory depends. Among them are the nearness in time of the item to be recalled, the frequency with which it has been recalled, how recently it has been recalled, the vividness of the memory, its relevance to the subject claiming attention at the moment, the number of associative links that tie it to other knowledge.

All this is self-evident. In terms of our metaphor the storekeeper knows at once exactly where to find memories that he has been handling recently and frequently, while those that have lain untouched on the shelves of the mind for a long time cause him some trouble. He has to search, as it were, in many places before he finds the right one. On such occasions 'quick as thought' is not so very quick.

In terms of another metaphor, often used by psychologists, frequently recalled memories reach consciousness by well worn and, therefore, unobstructed channels. There is, metaphorically speaking, little friction to the passage of the memories through such channels. Then the effort expended in forcing them through is below the threshold value.

The presentation of technical information

Associative links

The associative links that tie an item of memory to other knowledge guide, as it were, the storekeeper from item to item until he finds the required one. His search is not as blind as when the required item is detached. In other words, an item tied by associative links to many others has numerous personal implications and is the more readily recalled on this account.

Apply these considerations to a simple example. The task of explaining a new system of bidding at bridge has served us well already. So let this example serve us again. Suppose that you have occasion in the course of explanation to refer to some systems of bidding that are already in use. Plausible names for such systems might be the Philadelphia Team and the Four Knaves. At mention of the words Philadelphia Team and Four Knaves the storekeeper is sent off on a mission to collect associations. If John Smith has never heard of such systems the storekeeper will, of course, return empty handed, or, possibly, with associations that have nothing to do with bridge. If John Smith has heard of them but once, and long ago, the storekeeper will find his task difficult and it will take up an appreciable amount of his time. While he is looking for associations to the Philadelphia Team and the Four Knaves, while in other words John Smith is trying to remember what he once heard about these systems, the storekeeper cannot undertake another mission. If you expect him to, if after mentioning the Philadelphia Team and the Four Knaves, you immediately mention something else that causes a substantial

The work done by the person addressed

effort at associating, you will place upon the storekeeper a burden greater than he can carry. If, however, John Smith is quite familiar with the bidding systems of the Philadelphia Team and the Four Knaves he will recall all that he needs in a flash. *It is important always to consider, not only what John Smith knows, but also how well he knows it.*

The act of understanding

Let us now turn to the second of the acts performed by a person who is receiving new information, the act of understanding. This must follow on the heels of the act of associating. For understanding requires the correlation of the new information with what is known already. During correlation, work is done on the items that have been assembled for display to consciousness.

During the explanation of a new bidding system at bridge, for example, John Smith does not only have to recall the properties of the number 13; he also has to do some work on those properties, some simple mental arithmetic. He has to consider how the new bidding system would apply to hands in which the thirteen cards of a suit were distributed in some unusual way between the four hands, such as seven with one player and two with each of the others. He must also imagine the new system being used for the bidding of hands that he has recalled. He must effect a comparison between the new system and the systems of the Philadelphia Team and the Four Knaves. He must form in his mind a classification of bidding systems and fit the new one into this classification. He must, in more general terms, consider the implications of what he has just been told, form logical relationships between old and new knowledge, note similarities and contrasts between new and remembered facts, label, schedule, and classify the new knowledge in his mind. The act of understanding is, indeed, complex and varied.

The work of correlation, if done properly, is thorough. It results in a close association of old and new knowledge; so close that the one cannot be wholly isolated from the other. They form in the mind a single unit. They are welded into one, if I may be allowed yet another metaphor. In consequence what is put back on the shelves of the memory is inevitably somewhat different from what has been taken down.

Similar work was done to the old knowledge at the time of its acquisition. When long, long ago John Smith was being taught to count up to thirteen and learnt the simple rules of arithmetic, he was shown by his teachers how these rules are related to other things with which he was already familiar. Very likely he was taught to associate numbers with apples and oranges. For children are often taught arithmetic in this way. Without the act of associating John Smith could never have come to appreciate the significance of number.

It is the same with every one of the countless items stored on the shelves of John Smith's memory. Every one of them is connected by associative links with some other. The whole body of his knowledge really forms one indivisible unit in which no part can be very clearly isolated from the remainder. It is worth while to note

The work done by the person addressed

this if only because it shows that the storehouse for memories is, for many purposes, a misleading metaphor. It suggests too strongly a collection of detachable facts: but perhaps it is no more misleading than any other.

The act of memorising

As I shall have occasion to refer to the act of memorising in some detail later on I need say very little about it here.

In the metaphor that, for better or worse, I have been using, the storekeeper must, during an act of memorising, look for suitable places on the shelves of the mind and store the items to be remembered in those places. He must put them in position securely, or they will soon slip off their shelves and be lost. In terms of our metaphor the work of securing an item in position on the shelves of the memory consists in tying it to items already stored there. In other words, the work of memorising is intimately connected with the work of correlation to which I have just referred. When an item of information has become attached to a number of other items by associative links it has found a permanent place in the memory. An item can never be remembered in isolation.

To prove this let anyone try to commit to memory a short sentence in a foreign language unknown to him. He will find it very difficult. And the reason is the lack of associative links by which it can be secured to the shelves of the memory.

When trying to remember a sequence of sounds of which the meaning is not known one must, in the lack of natural associations, form artificial ones. With a little trouble one finds something of which the sounds remind one and associates them from then onwards with this. Such artificial associations are called mnemonics.

A mnemonic may be a jingling rhyme; as often used in the teaching of Latin grammar. It may be an easily remembered sentence in which some feature, perhaps the initial letter of each word, forms the necessary associative link with the thing to be remembered. 'A Cow Eats Grass' is an example. The letters ACEG designate the successive notes represented by spaces in the bass clef. A mnemonic of a different type is the well known Fleming's Rule in which the thumb, the first and the middle fingers point respectively in the direction of motion, flux, and current in a dynamo.

The occasions when such very artificial aids to memory are necessary are not numerous. I point them out because they are extreme instances that illustrate the type of effort that the act may demand. Between these occasions and those on which an item is remembered without perceptible effort there is a wide range. Just as a good expositor estimates correctly the time and effort needed for the acts of associating and understanding, so he must also estimate these needs correctly for the act of memorising. I propose to give a few hints as to how he can make this performance easier in Chapter IX.

VI The proper pace

Pace and timing in the entertainment world

This chapter ought really to be written by a music-hall comedian. For it is concerned with a subject in which he is more expert than anyone else, the art of timing. He judges to a nicety how long it takes his audience to see the point of each of his jokes. Some he fires off in quick succession, as from a machine gun. He knows that each will be grasped instantly, as soon as it has left his lips. He knows too that by his speed he is stimulating his audience to its greatest effort at attention. Each man among them is made to work for his entertainment. If he is not alert and active minded he will miss something. And as each sees the point as soon as the joke is presented he is gratified in some secret corner of his vanity by his own quick-wittedness.

Then, when after skilful handling the audience is keyed up to the highest pitch of mental activity, when every member has been turned into a delighted and efficient collaborator, then comes some more subtle joke. It calls for the association of ideas that are stored on remote shelves of the memory. Such a joke is not comparable to a machine-gun bullet, which hits its

The presentation of technical information

target instantly. It is more like a delayed-action bomb. It must be isolated from lesser, more obvious gags. So the astute comedian keeps the patter going with no more than a semblance of wit for just as long as need be till the arrival of that explosion of mirth, of that rumbling, spreading, contagious guffaw.

The great ones among music-hall comedians are sound psychologists. They know exactly how their audiences are occupied during that interval between gag and laugh. They know that the storekeeper is collecting together the associations that are called up by the comedian's words. They know too that it is not until these associations have been displayed before consciousness in all their startling incongruity that the energy of their high-explosive combination finds its release in laughter.

To know accurately to within a split second when an audience has reached the height of its receptivity to each quip and sally, to act with such complete assurance on his intuitive understanding of the mental processes of others as is done by the first-class music-hall comedian, reveals a degree of virtuosity only to be acquired after many years of hard practice and experience. It would, I fear, be unreasonable to expect so much meticulous care to the craftsmanship of presentation from every expositor. One rarely finds it elsewhere, even in the entertainment world. But though it may not be quite so important to their livelihood, straight actors, film producers, dramatists, novelists, essayists, all must give some thought to problems of timing. They must so adjust the pace at which they present their material that the person addressed has reached his maximum receptivity for each

The proper pace

point at the moment when it is presented to him. All those who take the expositor's art seriously know that the most dramatic situation, the lightest humour, the profoundest wit, the most illuminating revelation of truth, will fail to reach its mark if the proper pace has been misjudged.

In the theatre it is important for each character to appear on the stage and leave it at the moment when his entry and exit will be most effective, for each significant remark to be made at the most telling moment, for every situation to be prepared at the most appropriate pace; if it is too fast the audience will fail to notice something of importance; if it is too slow the audience will lose interest and receptivity. Similarly, in a novel the plot must be developed at the rate that comes naturally to a reader who would keep up with it. With an orator whose pace is too rapid the attention of the audience tires, while it flags just as much with one whose pace is too slow.

Are pace and timing less important to the presentation of technical information? A little less important perhaps, but surely still of considerable importance. If a scientific fact or argument is presented at a moment when the person addressed is thinking about something else it will fail to reach its mark. And if something fails to reach its mark in a learned society it certainly does not matter less than if something fails to do so in a music-hall. The attention of a person reading a scientific treatise is as liable to wander as the attention of a person reading a novel. A Fellow of the Royal Society is entitled to expect from a lecturer some help with the task of maintaining his receptivity.

The presentation of technical information

Yet he but rarely receives it. Most of those who have technical information or philosophical argument to present have done little towards solving their problems of pace and timing. One hardly ever hears them discuss these problems. I doubt if they know that there are any. So low has the standard of literary craftsmanship sunk in the scientific and philosophical worlds.

An example of too rapid a pace

Lest it be thought that for a simple statement of fact such problems do not arise let me give an example. 'The coal on arrival at the coal sheds is weighed and electrically tipped into a hopper and fed onto a conveyor, which carries it to the top of the boiler house, where it is tipped onto two drag conveyors, after magnetic extraction of any iron present, the coal being directed to any boiler at will.'

The above sentence only slightly caricatures a style from which we have all suffered many times. But what is wrong with it? The grammar and syntax are perfectly correct. It does not contain any unnecessary words or involved phrases. The language is simple and concise. It contains no recondite mathematics, no elusive argument, nothing but straight forward statements. And yet it imposes a great strain on the attention. The reason is that the pace at which the information is presented is greater than the pace at which it can be absorbed.

My example contains no less than seven significant statements. They are all crowded into one sentence and connected together by conjunctions and subsidiary

The proper pace

clauses. Each of the statements sends the storekeeper to a different set of shelves. 'The coal on arrival at the coal sheds is weighed.' This first of the statements requires that a picture be formed of the coal arriving from somewhere, of those buildings referred to as coal sheds, of the weighing operation. For the significance of the statement to be appreciated the John Smith to whom it is addressed must, moreover, think about the reason why the coal is weighed at that place. All this takes time, and before he can complete the work he is told that the coal is tipped into a hopper.

I think he might just be able to take notice of the fact if the word 'electrically' were not added. But this increases John Smith's effort enormously. What other methods of tipping might there be, he is expected to ask himself, and how do they compare with the electrical method? Such questions set the storekeeper off on a new mission. But no sooner has the stimulus been received than something else is mentioned, namely a conveyor.

From this picture the mind is asked to proceed to the top of the boiler house. Given time John Smith can, no doubt, remember well enough what a conveyor is like that moves coal over a great vertical distance. A second or two would suffice to impress this stage in the coal handling operation on his mind. But he is not allowed a second or two. Instead he is asked to picture another type of conveyor, called a drag conveyor, only to surrender this picture for something even more remote, namely magnetic extraction.

And so it goes on. If John Smith reads this passage only once he will certainly not notice that it contains

The presentation of technical information

seven significant statements. He will miss all but two or three of them. The simple reason is that as his eyes peruse one thing his mind is preoccupied with something that has gone before.

The young scientist is sometimes told, a little pontifically, that he must be as concise as he can; that he must never, never use two words if one suffices to express his meaning. The above example shows that such advice may be very misleading. It arises from a confusion between what is needed to *express* a meaning and what is needed to *convey* it. Tell the young scientist by all means not to use more words than are needed to convey his meaning. But if he is restricted austerely to the bare number needed to express it he may commit the error of overcrowding, which is just as bad an error as that of verbosity. If the John Smith addressed fails to notice the one word that so concisely expresses an author's meaning that one word might just as well never have been uttered. If no less than a hundred words are needed to ensure receptivity a hundred words must be used. They are all functional. The scientist who prides himself on his concise laconic style is often a mere soliloquiser. He is too much interested in his own achievements, not enough in the John Smith whom he is addressing. To such a scientist technical information is something to be placed on record, not something to be conveyed from mind to mind. The rule is: *Add to the words needed to express the information as many more as may be needed to convey it.*

The proper pace

Further examples

A few further examples of overcrowding may help to illustrate this rule. They are not caricatures but the real thing. I have read each one of them somewhere and copied it out because it is typical.

'The pulverised fuel is gravity fed to a point where it meets the forced draught through a twelve armed spider rotating rapidly to break up any lumps of coal that may have formed due to the coal particles adhering together.' It would not help the author of that sentence to tell him that he must be concise. He should be taught to force some extra words and a few full stops into his sentences as a sailor forces a marlinspike into a tightly twined rope. Such treatment would loosen it up and make it comprehensible.

'The transformers are water cooled and there is a flooding system in the building so that, in the event of fire, the whole place is automatically flooded.' I doubt very much whether the fact will impress itself on the mind of those who read this sentence that the transformers are water cooled. The interesting information about fire risks, which follows closely on its heels, comes too soon. It will eradicate any impression made by the first five words. The storekeeper will be stopped as soon as he has begun to search for associations to the various available methods of cooling transformers and sent off on a quest for more exciting memories about fires. If the method of cooling the transformers is important enough to merit notice it should have a sentence to itself together with enough time to allow the information to sink in.

The presentation of technical information

'Although the station is owned by the Corporation all its running costs and capital charges are paid by the Central Electricity Board, who own the power generated, and pass on telephonic instructions to the control room, when extra power is required in the Grid.' Here the fault of overcrowding is made worse by bringing together items that are logically unconnected, or only loosely connected. Questions of finance and ownership are mentioned in the first part of the sentence. Questions of control and instructions in the second part. They are distinct subjects and ought to be discussed under distinct headings, not in one sentence. Overcrowding is always bad, and it is worst when the things that are herded together have incompatible natures. It is not uncommon to find a combination of the two faults. A writer who packs his sentences regardless of their holding capacity packs hastily. Such a writer does not examine the items very carefully. He does not pause to ensure that statements that follow each other closely shall be connected logically. In fact his lack of sense of pace is probably accompanied by a lack of sense of logic. The rule is this: *In the spacing of successive items consideration must be given to their logical connection.*

There are other occasions when overcrowding does not arise from the inclusion of too many detached items but from presenting a single item with too much detail. 'As the gases leave the furnace they heat the incoming air to 400°C and pass into a Lodge Cotterell electrostatic precipitator.' If here the words 'to 400°C' were omitted the sentence could be taken in without effort. A simple fact such as that the gases heat the incoming air is easily

The proper pace

understood while the words are being read. The storekeeper will be ready to collect associations to the electrostatic precipitator by the time the next statement has been reached. But a number calls for more work than a simple qualitative statement. 400°C has to be appreciated and remembered independently of the bare fact that the gases heat the incoming air. The John Smith who reads that number must pause to consider whether it is a large or a small one, whether he should expect it, what its implications are. Then he must memorise it. And numbers are not as easy to remember as simple facts. John Smith must stop reading after he has come to 400°C while he does the necessary work on it.

Now stopping and starting is as energy consuming in reading as it is in motoring. A good author provides a clear run. He would have done so in this instance if he had written: 'As the gases leave the furnace they heat the incoming air to 400°C. After they have done this they pass into a Lodge Cotterell electrostatic precipitator.' This provides a separate sentence for each piece of information. It allows, while the five words 'after they have done this' are being read, sufficient time for the work of understanding and memorising the number 400°C. These extra five words, together with the full stop, show that there is another statement to follow. The storekeeper is warned, as it were, to hold himself in readiness for a new quest. Here the rule is: *So arrange your discourse that numerical information is clearly isolated from adjoining items.*

Errors of pace are obviously more serious in a lecturer than in a writer. An audience in the lecture room has no control at all over the rate at which it receives the information. It cannot stop the lecturer or ask him to repeat an earlier statement. It is entirely at his mercy. The reader of printed matter, on the other hand, can put the book down from time to time and ponder on what he has just read. He can turn back the pages. He can read a difficult passage over and over again. He can skip when he comes to a stretch of arid verbosity.

Reading as an automatic process

However, John Smith does not always have such complete control over the pace of his reading as authors seem to assume. He does have complete control when he is using a telephone directory or a railway guide. He has enough control to compensate for an author's bad presentation when he is consulting a work of reference and needs only to peruse a short passage at one time. But he has little control over the pace of his reading when he is perusing a lengthy description or argument. The reason is purely psychological.

For most of us, reading has become a very automatic process. The eyes can, and often do, read on while the mind is preoccupied with something else. This is what is likely to happen to John Smith if you either bore him by keeping the pace too slow or weary and confuse him by keeping it too quick. The reading habit is too strong for him. Though you may hope that he will stop and collect his thoughts whenever they have begun to wander, though you may consider it his duty to put the printed page aside until his receptivity has been restored, you must not count on his being so obliging. If you do you are not a persuasive writer.

The importance of being stimulating

But you cannot become one by diluting your discourse with meaningless words. Whatever you add must serve to increase John Smith's alertness at the time when the information is being presented. I have said already that this is largely under your control. A characteristic of human psychology is that a person is receptive when he is stimulated and loses receptivity when he is bored. So a good expositor does what a good music-hall comedian does. He takes care that everything he says shall have meaning. So long as he succeeds he can safely maintain a rapid pace. As soon as he fails John Smith loses receptivity altogether and no slowing down of the pace will restore it. Hence another rule: *Anything introduced to give the person addressed time for the work that he has to do should be of such a nature that it keeps his mind occupied.* To do this it must tell him something; it must have meaning.

This sounds paradoxical. For it seems to suggest that if your discourse is too crowded you ought to crowd it still more by squeezing in additional information, that if John Smith has more work to do than he can accomplish in the time, you must give him still more work. But it is not so. The things with which you keep his mind occupied must be the things with which it has to be occupied anyhow in order that he may receive the information that you are imparting. The things that you tell him

The presentation of technical information

must be things that will help with the work that he has to do, not things that add to that work.

The work consists, it will be remembered, of three acts: associating, understanding, and memorising. The first act is helped when John Smith is told what the next item will be about. The second act is helped when he is told what the information just provided means. The third may be helped in a variety of ways that are not easy to describe in a few words. Those who present technical information fail frequently to give sufficient help with these three acts. So I think it will be useful to discuss, in some little detail, in the next few chapters how this may be done.

VII What it is about

Presenting a simple instruction

Suppose John Smith is a rather slow-witted labourer and you are the works manager. You meet him in the factory yard and say to him: 'Tell Freddy Butler I want him in my office.' Is the form of words functional?

It is simple, certainly, and clear. It is admirably brief. If simplicity, clarity, and terseness were the only criteria of a functional style there could be no doubt that it was perfectly good functional English to say: 'Tell Freddy Butler I want him in my office.' By such criteria this form of words could not be improved. But the proof of the pudding is, as the saying goes, in the eating. Is this simple, clear, brief form of words the best to ensure that John Smith, slow-witted labourer, will do what you tell him to? If it has the effect, if its meaning is instantly as clear to John Smith as it is to yourself, then I have no criticism to offer. But if, on hearing these words, John Smith hesitates, if he is left uncertain as to what you want of him, then I shall have to accuse you of using unfunctional language.

Which it will be, depends, as it so often does, on circumstances. If you are in the habit of employing this

The presentation of technical information

John Smith for similar errands and if, in particular, you often ask him to send Freddy Butler to you, he will act on your words with alertness. But if your request is an unusual one, if John Smith does not have much to do with Freddy Butler, if he would hardly expect you to want that person in your office, then your brief words may be too precipitate. It would be better to expand them.

After all, to a slow-witted person your request must seem quite complicated, too complicated to be grasped while he is hearing a short sentence of only nine words. He has to bring several ideas into association. First he must recall who Freddy Butler is. Then he must think out where to find him. Then he must make a mental note of the message that he has to deliver. It is in two parts. First there is the person who wants Freddy Butler; it is the boss. Then there is the place where he is wanted; it is the boss's office.

John Smith is not an idiot boy. He is quite capable of understanding your request and acting on it if he is given time enough. But while you are saying those nine words all sorts of things seem to be crowding in on him at once. The boss, who is saying something; someone called Freddy Butler; someone wanted; something to do with an office. What is the connection between all these things? So John Smith stands and stares at you with a puzzled, vacant look on his dull face.

You save John Smith from distress and yourself from irritation if you arrange your words in such a way that each component part of the request reaches him at the moment when he is receptive for it. To do so you must

What it is about

introduce some punctuation signs and additional words with a helpful meaning.

'You know Freddy Butler. Tell him I want him in my office.' This might serve. It contains one new full stop and twelve words in place of the nine in the first formulation. There is a little gain in time and a bigger gain in stimulation. The storekeeper is helped with one, at least, of the various things that he has to do before the meaning of your request can be comprehended. This is to display an image of Freddy Butler in the conscious compartment of John Smith's mind. The words 'you know Freddy Butler' direct him straight to the right shelf. The storekeeper has completed that part of his work before your request is made. These words serve the sole purpose of saying what the coming information is to be about.

They may not be enough. If John Smith is *very* slow-witted you may have to allow a little time to elapse after having said 'you know Freddy Butler' before you proceed to say what you want. You will be able to see from the expression on his face when the storekeeper has completed his work and duly produced the image of Freddy Butler. You can shorten this time by adding some further words of helpful meaning. 'The fellow who works in the pattern shop', may serve; or 'he wears a striped jumper'. With such reminders you do not give any new information. You only tell John Smith things that he knows perfectly well already. That is why they lighten his work.

More than the above additions may be necessary. John Smith is to make a mental note of who is making the request, to whose office Freddy Butler is to be sent.

The presentation of technical information

To do this the storekeeper must bring an image of you into association with the image of Freddy Butler. To hail John Smith with 'you know Freddy Butler' may be too abrupt. It may only surprise and confuse him. If so it will be better to let him know who is speaking before you even begin on your request. You might open with 'Good afternoon, Mr. Smith' or, if you prefer it, with 'Hi, you'. Either form of words would be functional.

Then again 'tell him I want him in my office' may be too condensed for a very slow-witted person. It may be better first to say 'find him'. These words direct the storekeeper's steps to the set of shelves that contain knowledge of the places where Freddy Butler is likely to be found. Pictures come into the conscious compartment of John Smith's mind of the pattern shop, where Freddy Butler works, the canteen, where he eats, the secluded corner, where he enjoys an occasional clandestine smoke.

Thus there may be circumstances when a functional statement may have to be much longer than the nine words that suffice to express your meaning. It may have to be: 'Hi, you. You know Freddy Butler. He works in the pattern shop. Find him and tell him I want him in my office.' Twenty-three words, a comma and four full stops. All but the last eight words are designed to tell John Smith what your request will be about. You do so in stages so that the various sets of associations that must be collected may be well separated. First you make it clear by 'Hi, you' that it is yourself who wants something. Then you stimulate the effort to bring the image of Freddy Butler to consciousness. Finally you cause John Smith to think about the places where this person might

What it is about

be. As a result the important thing that you have to say, 'tell him I want him in my office', falls on an alert mind.

Presenting a comment in a committee

Such is the help that must be given to a slow-witted John Smith, labourer, when he is receiving a simple instruction. A quick-witted John Smith, F.R.S., needs exactly the same kind of help when he is receiving a piece of recondite technical information. But he does not always get enough of it.

Suppose that in a technical discussion you put forward a new and unexpected point of view. It can only be appreciated with the help of a set of associations quite different from those that are in the minds of those present. If you make your point too precipitately there will not be time for the storekeeper to collect the necessary associations before he is stimulated to a different mission by a remark interposed by someone else. How often it happens at a committee that some shrewd person's wise and important comment passes unnoticed merely because he has failed to preface it with the equivalent of 'You know Freddy Butler'.

Prefatory statements

Similar prefatory remarks are needed for the information that is being given in a paper or a book. Authors, it is true, do not neglect this duty altogether. They usually indicate by title and subtitle what the paper or book is about. They open, perhaps, with a clear statement of the existing state of knowledge on the subject to be discussed. They may even tell the reader in a preface something about the ground they propose to cover and the method they propose to adopt. In many offices the first page of every technical report must, as a matter of routine, contain a summary of the terms of reference. Such introductions are very necessary. But the help they give is not enough. More, much more, is needed than a single prefatory statement at the very outset of your discourse.

The contents of the conscious compartment of John Smith's mind have to be changed many times during even a brief discourse. Each new chapter, each new section, each new paragraph, to some extent each new sentence, brings a change of subject. Every new subject imposes on the storekeeper the duty of collecting new associations. Whenever this happens alertness in the person addressed must be re-created, the storekeeper's steps must be guided afresh. When the change is of any magnitude this can only be done by the introduction of words to say what the next item will be about.

So the same rule that applies to a single isolated remark applies also whenever a new subject occurs in the body of your discourse. Whichever it be you must always make sure that John Smith knows what it is about at the moment when it is presented to him. Here again functional writing differs from imaginative writing. An imaginative author will sometimes deliberately keep his reader mystified for a while. He allows it to appear only gradually who the people are who are speaking, what they are talking about, in what situation they find them-

What it is about

selves. It enhances the pleasure of a John Smith who is perusing imaginative literature to find himself first puzzled and then enlightened. He is stimulated by the work he has to do while he is piecing together the clues by which the author lets him discover, bit by bit, what it is about. But when your discourse is technical John Smith cannot afford this effort. If there is a risk that he will not know in good time what it is about, you must tell him. Your discourse must be plentifully interspersed with prefatory statements.

A prefatory statement may serve a variety of purposes. Let me list what seems to me some of the most important ones:

(a) to define the situation
(b) to define the theme
(c) to be a reminder of existing knowledge
(d) to be a subsidiary reminder
(e) to define the method

Sometimes a single prefatory statement may serve more than one of these purposes.

Statements of the situation

The situation is defined in the example given above, very unconventionally, by the words 'Hi, you'. They draw attention to your presence in the factory yard. A more hackneyed form of words for defining the situation is: 'Referring to your letter of the...' These introductory words make it clear that the situation is the need for a reply to a specific letter.

Other forms of words that may serve the same purpose are: 'You asked me yesterday to report to you on ... ', 'We have met, gentlemen, in order that we might consider ... ', 'I have received a request from Mr. X that I should explain to you ... ', 'May I remind you of our recent conversation when I promised to find out about ... ', 'Some doubt has been expressed as to whether the strength of the tie bar in this structure is of the right value. So I have made some calculations and find ... ', 'While the effect of circular cross sections has been thoroughly studied that of rectangular ones has hitherto been neglected'. The need for such statements is obvious and yet many careless or clumsy authors leave them out. The effect on the reader of their omission is to give him a nasty jolt as when a train accelerates too rapidly.

Statements of theme

Some of these examples contain a definition of the theme as well as one of the situation. Other forms of words serve to state the theme only. 'Let us now give our attention to so-and-so', 'Referring to the tie bar', 'In so far as rectangular cross sections are concerned'. These are all too stereotyped forms. All I can say for them is that they are better than nothing. But their repeated use becomes monotonous. A good expositor finds more varied, more subtle, more stimulating ways of announcing the new theme.

One of them is very simple. If the storekeeper does not need a great deal of help it may suffice to arrange the opening sentence of the new subjects so that the word to guide the storekeeper's steps occurs early in the sentence. Let me invent an example. You are discussing a certain structure and propose to say something about the strength of a tie bar in it. You have not yet mentioned the tie bar. You might begin a new paragraph with the words: 'A careful calculation has been made and shows that the tie bar is strong enough'. John Smith has to read nine words before he knows that you are referring to the tie bar. In the meantime the storekeeper is left standing about idle. Alternatively you might open with: 'Referring next to the tie bar . . . ' This may be overdoing it. A happy mean would be to say: 'The tie bar has been calculated and is found to be strong enough'. Placing 'tie bar', which here serves as a theme word, near the beginning of the sentence enables the storekeeper to complete his task just in time.

Sometimes, of course, a 'theme word' is not enough. If the change of subject is radical a 'theme sentence' may be required. On some occasions a whole 'theme paragraph' or more is needed in order to define the theme of the next part of a discourse. When this happens the form chosen for an announcement of theme deserves considerable care. I have said already that hackneyed methods should be avoided. John Smith associates to the new theme more readily if he is stimulated by it. So an author should make his announcements of theme as interesting as he can. The task provides scope for all his skill and inventiveness. Indeed, nothing is more individual to an author, nothing characterises his style more surely, than the methods he uses for statements of theme.

The presentation of technical information

One of them, the appropriate quotation 'Darwin said so-and-so ...', is typical. One introduces a subject by mentioning first what a well known authority has said about it. There are endless varieties of this method. In one of his famous prefaces Bernard Shaw wants to introduce economic considerations into his discussion of the London theatre. He might have said: 'Turning now to economic considerations'. But what he does say is: 'Wellington said that an army moves on its belly. So does the London theatre. Before a man acts he must eat'. That is but one of the possible ways of doing it. Another author would, perhaps, have cast his statement of theme in the form of a quotation from Shakespeare, or from *Alice*, or from scripture. Another might have chosen to begin by telling an anecdote the point of which was that economic considerations play their part in human affairs as well as romantic ones.

There are occasions when it is perfectly functional for a scientist to use a similar technique. During a debate or at a committee meeting the quotation, the anecdote, the joke are often the most effective means of introducing a new subject and conveying information on it from mind to mind. In a paper to a learned society the quotation and the anecdote must be used more discreetly. But even there they are sometimes justified. A quite unconventional method of stating the theme is good if it directs the storekeeper's steps to the right shelves and to no others. But to do so it must be very strictly relevant. If not it diverts his steps to shelves where associations are stored that are alien to the theme, and then the method ceases to be functional.

What it is about

Reminders

A reminder of existing knowledge is useful whenever a mere statement of theme does not suffice to guide the storekeeper's steps. In our example it took the form: 'You know Freddy Butler'. In the presentation of technical information the form might be: 'It is well known that...', 'The formula that applies here is, of course...', 'You will remember that...' Such reminders are superfluous if John Smith can recall his existing knowledge with perfect ease. But if the act of recalling causes him the least effort, then you should spare him that effort by a reminder.

Subsidiary reminders

A single reminder may, moreover, not be nearly enough. And this is why in the list given above I have mentioned a subsidiary reminder separately. I want to point out that prefatory statements often need to be very elaborate. In the example given at the beginning of this chapter a subsidiary reminder was: 'He wears a striped jumper'. Here is another example: 'These small creatures, it will be remembered, were mentioned by X in his famous monograph. They are often observable in rockpools in early summer.' The first sentence is a reminder and the second a subsidiary reminder. Or: 'Rectangular cross sections have recently been the subject of controversy' combines statement of situation, statement of theme, and reminder. 'They occur in our B6 model' could follow as a subsidiary reminder. If a subject is to be

The presentation of technical information

introduced that is very unfamiliar to the particular John Smith addressed, a whole string of subsidiary reminders may be needed. When lecturing to a class of students it is sometimes worth while spending ten minutes on recapitulation of what they know already, in order to help them to grasp one new piece of rather recondite information.

Experts in their own subject need, on the other hand, few reminders, if any. And when you are presenting technical information you are probably doing it as an expert. You need therefore fewer reminders than the John Smith whom you are addressing. Remember this. Do not expect John Smith to grasp what it is about as quickly as you yourself can. In your choice of reminders and subsidiary reminders be guided by what John Smith needs, not by what you need. This advice is by no means superfluous, for it is far commoner to find too few prefatory statements in technical literature than too many.

Statements of method

By these I mean statements that say, not so much what the next subject will be, as how it will be treated. Such statements are common enough in prefaces and introductory chapters. In these an author will tell his readers whether the subject matter is to be advanced or elementary, whether the approach to it is to be on new or on familiar lines, whether it is to be considered quantitatively or qualitatively, whether a philosophical or a scientific point of view is to be adopted, whether practical or

What it is about

theoretical aspects of the subject are to be considered, whether the treatment is to be in terms of this or that branch of mathematics. Without advance knowledge of this type John Smith may be puzzled by a method of treatment that he is not expecting.

However, a broad statement of method at the very beginning of your discourse is not enough. At any moment the situation may arise where John Smith expects one of the alternatives listed in the preceding paragraph. If you give him the other he will be puzzled and lose receptivity. So the type of prefatory remark that I am calling a statement of method should occur frequently in the body of your discourse.

Here are a few examples. 'I must now dwell in some length on so-and-so.' Or: 'This subject must, however, have a chapter to itself'. Such remarks tell John Smith that he must settle down to a prolonged consideration of the next subject. On the other hand, 'Let me point out in passing...' tells him that the storekeeper must not yet clear away everything that has been accumulated for consideration of the main subject, for you will soon return to it.

There are other occasions, too, when it is useful to warn the storekeeper of the nature of the work that he will have to do. 'A matter of minor importance is that...' This allows the storekeeper to save his strength for an occasion when a matter of major importance will be presented. Such an occasion is correspondingly announced by such words as 'Last, not least', 'It would hardly be possible to exaggerate the importance of...'

On other occasions, again, it is useful to let John Smith

know whether you are at the moment engaged on presenting a series of items that all belong to the same category or to different ones. If, for instance, you are about to say something that supports what has just gone before you may use the word 'moreover' as a statement of method: 'Practical considerations, moreover, lead to the same conclusion'. The corresponding word to show that the next item does not belong to the supporting but to the contrasting category is 'however': 'Practical considerations, however, lead to such-and-such a conclusion'. Without the word 'however' John Smith might not be quite sure whether such-and-such a conclusion did or did not contrast with the former one. 'On the one hand', 'on the other hand', are similar forms of words that I would describe as conveying a statement of method. They tell John Smith that you are making a comparison.

There are further occasions when a statement of method cannot be very sharply distinguished from a statement of theme: Yet there is some difference. For the statement does not direct the storekeeper to the precise shelf where a given association will be found. It only indicates the general direction in which he has to go. This would be done, for instance, by: 'Purely deductive reasoning leads one to conclude...' John Smith is prepared in advance for deductive reasoning. Or: 'Practical considerations show that...', 'No doubt a philosopher would point out that...', 'Expressed qualitatively only...', 'In mathematical terms this can be expressed as follows', 'In the language of quantum mechanics one can say that...' All such statements help

What it is about

John Smith to adjust his mind not only to the subject but also to the way in which the subject is to be considered.

Similar help can be given when an enumeration is being presented. If you have to mention two things you will often be well advised to say beforehand that you are going to mention two. If you propose to enumerate four reasons in support of your recommendation, say to begin with that there will be four reasons. Then John Smith can tick them all off on his fingers, as it were, as he comes to them and he will know when the end has been reached and he may expect to be told about something else. Sometimes it is even useful to let John Smith know approximately how much space you will devote to each of the enumerated items. He can perform the three acts of associating, understanding, and memorising, better if he knows how much time he will have for each.

However, there is no need to say in so many words how much space will be given to each item. You have at your disposal subtler, less obtrusive means of doing it. Among them are paragraphing, punctuation, the numbering or lettering of paragraphs, marginal headings, sub-headings. They are all visual aids that tell something about your method of presentation.

Perhaps it is worth while mentioning yet another way of doing it. It is sometimes functional to introduce each item of an enumeration with an identical form of words. It is the technique of the litany and conveys to the ear what a new paragraph conveys to the eye. I do not recommend its use, except very sparingly and discreetly, for the presentation of technical information. But I do

Denials of intention

Among these statements I would like to list just one more type. I will call it a denial of intention. By this I mean a statement that says what it will *not* be about. It should be given whenever some evocative word may otherwise send the storekeeper off in the wrong direction. If he brings a number of unwanted associations into the conscious compartment of John Smith's mind there is bound to be trouble. Before they have been put back on their shelves a permanent misunderstanding may have come about.

Suppose, to take a rather extreme example, that you have occasion to say rather unexpectedly something about chrome and that chrome happens recently to have been the subject of political controversy, as well may be. Anything may become the subject of political controversy, from cheese to page's buttons. Then there is a risk that John Smith may think for a moment that you are about to introduce politics. The new associations that crowd into his mind with this thought may cause him to miss the point of your next remark. It is on such occasions that a denial of intention is useful. One of your prefatory remarks should be: 'What I am about to say has nothing to do with politics.'

Similar false associations have their disturbing effect in less obvious instances and when the associations are more closely allied to the subject of your discourse. You

What it is about

may, for instance, be about to make a purely qualitative statement when John Smith has some reason to expect a quantitative one. Then a denial of intention may take the form: 'Without going into figures', or 'Speaking purely qualitatively'. Other denials of intention are contained in such forms of words as: 'Relativity considerations apart', 'Let us limit ourselves to the cases where x is small', 'I shall not attempt a high degree of accuracy in this', 'I do not propose to say anything in support of the theory of Mr. X'.

If, moreover, you are about to present some fact that is difficult to accept when John Smith might expect something quite easy, you may introduce a similar denial of intention with the words: 'Though it is not quite obvious a little thought will show that . . . ' If it is the other way about you may say: 'No great effort is needed to realise that . . . ' *The general rule is that a denial of intention should be introduced whenever there is the least danger that John Smith's receptivity may be impaired by a misapprehension.*

Transitions

All the four types of prefatory statement that I have been discussing occur at a moment when a transition is being made from one subject to the next. And this observation leads naturally to a consideration of transitions in general. An important thing in their management is that they should always be clearly recognisable as such. John Smith must not be left in any doubt that you are leaving one subject and passing on to another one. For the storekeeper has to clear away from the conscious compartment

The presentation of technical information

of John Smith's mind what is no longer wanted there as well as to bring into that compartment what is wanted. To enable him to do this at the right moment he needs a recognisable sign to show that a transition is being made.

With a skilful author such signs occur at every turn. They are revealed by the structure of his sentences, by his paragraphing, his punctuation, the way he puts things. He does not only know how to plan his discourse but also how to reveal the plan to John Smith. If need be he announces the plan with the help of the lettering and numbering of sections, with headings and sub-headings.

Many of these devices appeal only to the eye. So they are of no avail for the spoken word. Changes of tone, gestures, long and short pauses must largely take their place. It is really surprising how few scientific lecturers seem to know this. They pass from subject to subject only too often in a perfectly level voice, without a pause or gesture to indicate to their audience that a transition is being made. And those who hear them have the greatest difficulty in keeping awake and alert. But why should they struggle? Even if they strain their attention to the utmost they can still not discover when one subject has been disposed of and another started, so they are at a loss to know what much of the lecture is about. The most cooperative audience must miss much of a lecture in which transitions from subject to subject are not clearly marked. This is why a bad lecturer conveys much less than the written word and a good lecturer can convey so much more.

What it is about

Importance of architecture

Now one of the most important rules about prefatory statements. *Always keep your promises.* Having told John Smith what it will be about make quite sure that it is really about what you said and not also about something else.

This rule would be easy enough to follow if one could always plan one's discourse down to the minutest detail before setting pen to paper. But that is difficult even for a straightforward narrative. And when the discourse contains a certain amount of reasoned argument it is very difficult. For most of us do some thinking as we write. Indeed, we must sometimes write in order to find out what we think. It is therefore inevitable that some of your brightest thoughts should occur to us out of their proper turn. While we are engaged on Section B we think of something that rightly belongs to Section A. While we are penning one sentence something comes to mind that would better have been said just before the last sentence but one.

It happens so often like that. During a discussion of boiler-house control something comes to mind that is more apposite to the work of a switchboard attendant. When a list of disadvantages follows a list of advantages it is suddenly realised that one of the disadvantages in the list is in certain circumstances more of an advantage. One ought to have said so while compiling the list of advantages. While discussing the theory of X, which has been announced in a statement of theme, one remembers that the theory of Y, though not announced, is similar and ought to be discussed too. When one has just mentioned a number of desirable changes in the organisation of one's department one thinks of a desirable change that has been left out. These are but a few of the innumerable kinds of afterthought that may come to any author while he is writing.

Though scientists have minds that have been trained and should be orderly, they set down such bright afterthoughts as and when they occur far more often than one would believe possible. The result is a most confused unreadable style. Any remark that is allowed to slip into your discourse unannounced finds John Smith unprepared for it; he has no associations ready. So he will probably miss it, no matter how important and valuable it may be: in his puzzlement he may also miss other items that *have* been announced. In fact, any item in the wrong place is a foreign body; and, like all foreign bodies, it does harm there. Not only does it fail to achieve its purpose; it also prevents other items from achieving theirs.

An author who is full of ideas is obviously more liable to get some of them in the wrong place than a dull author. The probability that everything he has to say will occur to his active mind just at the precise moment when it is most suitably presented to John Smith is not great. So an active-minded writer has to be particularly careful about the proper placing of every item. He is, unfortunately, often the last person to notice that something has crept in unannounced. He himself has all the necessary associations to it ready to hand as he is writing. It is therefore particularly difficult for him to realise that John Smith will not have the same associations while he

What it is about

is reading, that the storekeeper will have a great deal of fetching and carrying, of packing and unpacking to do unless he, the author, does a great deal of rewriting.

This necessary rewriting must sometimes be very thorough. It is a most irksome duty and those who have technical information to present cannot easily afford the time and labour that it entails. But then, let me add, neither can they afford *not* to rewrite. If they allow their discourse to reach the public badly architected their best thoughts will pass largely unheeded.

There are ways of dodging this irksome obligation, I know. An accomplished and facile writer has learnt in time how to get away with a lot of bad architecting. He discovers how to make his presentation so smooth and attractive that his many unannounced changes of subject pass unnoticed. His wit and inventiveness serve to divert attention from his lack of logical consistency. He manages somehow to weave each bright thought into his discourse as and when it occurs to him, and so neatly that its irrelevance in that place is not too apparent.

Such accomplishment is rare among true scientists and found more often among those who dwell on the borderlands of science and mix a little ill-digested philosophy with some popularisation of science. They are widely read for the pleasure that they give at the moment. Their attractive style is much praised and it probably deserves the epithet scintillating. But it does not deserve the epithet functional. For it lacks what is essential to a functional style: logic and discipline. Hence it has little lasting effect. The John Smiths who read it with enjoyment do not retain what they have been told for any longer time than they retain the plot of a thriller. No. There is no effective alternative to good architecting. If John Smith is to have a sufficiently clear knowledge of what every item in your discourse is about, to fit it into an integrated whole and understand it in its relation to your main theme, you will have to take enormous pains with problems of architecting.

VIII Making it easy to understand

How an imaginative author conveys his meaning

Perhaps the biggest difference between an imaginative and a functional style is that, in the former, one should often allow facts to speak for themselves, while, in the latter, one should rarely do so.

Shakespeare does not say explicitly that King Lear is, among other things, an obstinate, obtuse, vain old man, although these characteristics are of decisive importance; for there would be no play if King Lear's qualities did not include these defects. Nor does he tell us in so many words that Richard II is a vacillating monarch, weak and boastful, who uses the flimsy prop of rhetoric to shore up his crumbling courage. Nor again, does he tell us that Othello possesses, together with his nobility and animal impulsiveness, a rich exuberance of spirit that could only have been nurtured under the fierce African sun. Although Othello's actions cannot be comprehended without appreciation of that exuberance, we have to infer it from the imagery contained in his speech. Shakespeare's plays bring something significant in every word, but we are not told by the author what the significance is. We have to discover it for ourselves.

The presentation of technical information

Novelists expect us to do the same. The more skilful they are the less do they explain and the more do they allow to be inferred, the more do they rely on the insight and sensitiveness of their readers, on their sympathetic understanding of hidden motives and subtle relationships. This is true, to a limited extent, even for the lowest type of thriller. When the villain has locked the heroine in a dungeon and strangled her dear white-haired old mother, the author does not tell us that he is a bad man. The crudest literary hack would feel that it was inartistic to be so explicit. In better work an author will describe scenes and sounds. He will tell us, perhaps, in minute detail what the heroine looks like. His means of conveying the appearance of things are often plain, straightforward description. But he does not use it much for other qualities. An author only tells his readers that the heroine is witty if he cannot invent witty things for her to say. If he can, he allows her wit to be revealed by her conversation.

A clever author finds a dozen ways of suggesting the subtle reactions that people produce in each other. And yet he never says in so many words what the reactions are. Maybe he does not even comment on them. Though the reasons that prompt men's conduct may be the centre of interest in his novel he does not necessarily say one word about those reasons. He prefers that the reader should guess at them. Whenever events are able to speak for themselves the cunning author lets them do it. The beauty of imaginative writing consists largely in what is *not* said.

The same rule applies to the technique of imaginative films. Much of the story is conveyed by inference. A hand tentatively advanced and then quickly withdrawn;

a crumpled scrap of paper blown across a deserted railway platform; the sound of a dog whimpering behind the door to a flat; with the help of such detail an inventive producer builds up the story. A substantial part of what is supposed to happen is not shown on the screen; it is provided by the imagination of the audience. So it should be. For the pleasure that one gets from imaginative art is the pleasure of exercising one's imagination, one's insight, one's understanding of human nature. In such exercise lies most of the work done by the person addressed in imaginative writing.

A great writer, a Shakespeare, provides so much for this exercise that the spring is never exhausted. Today, after three and a half centuries, people are still discussing what Shakespeare meant; they are finding new depth in his work. It has been asked whether Shakespeare meant Falstaff to be likable or unlikable, comic or pathetic; what Hamlet really felt for Ophelia; what twist caused Henry IV to connive at murder; whether Shylock's creator meant to portray a grasping rogue or a tragic representative of the Jewish race, or whether he was primarily concerned with providing a fat character part for some talented actor of the Globe Company. All these interpretations can be justified: so can many others. Subtler and deeper ones. Therein, people say rightly, lies Shakespeare's greatness.

The need for explicitness in functional writing

How different it is with functional writing. There cannot be two different opinions as to what Newton

The presentation of technical information

meant to say. One does not measure a scientist's greatness by the amount of work that scholars need do in order to interpret him to the world. If any doubt were left as to a scientist's meaning after enough time had been allowed to study him, that would not prove him to be a great scientist. It would only prove him to be a poor expositor. A scientist must not leave anything to the imagination. He must set it all down in black and white, as Newton did. The scientist must explain where the poet implies.

And what the scientist must do everyone must do whenever he is using the style that I call functional English. Though it is on a lower level to explain than to imply, he must dwell on that lower level. You, who are presenting new information to John Smith, must remember that he has other work to do while he is attending to you than while he is attending to an imaginative author. Like this latter, you tax a set of his mental faculties, but you tax a different set.

The act of associating may be much the same whether John Smith is being told what is the result of introspection or of inspection. But the act of memorising is usually more difficult in the latter case and the act of understanding is of quite a different order. If I may be permitted a little over-simplification I should say that imaginative writing calls for insight where functional writing calls for reason. At any moment when you are expecting John Smith to use his inductive and deductive powers you must not expect him to use his imagination as well. If you are to help him with the act of understanding you must guide him to every conclusion that you wish him to reach.

Making it easy to understand

We all fail to do so at times; particularly when the conclusion is very obvious to us and should also be quite obvious to John Smith. Then we tend to think that it is redundant to state the obvious. But quite often we ought to do so all the same. Let me invent an example.

A girder has been designed in your firm and John Smith, who is your boss, has asked you to report on it. In your report you discuss the strength of a certain tie bar and sum up the result of your calculations with the words: 'The stress in the tie bar is 25 megapascals'. I venture to suggest that as a final conclusion this statement is not functional.

John Smith wants to know whether the design ought to be changed or not. To reach a decision he needs the factor of safety in each member. In this instance, let it be supposed, the breaking strength of the material is known to be 500 megapascals. So the factor of safety is 20. This is obvious. As obvious, indeed, as that the villain in a thriller is a bad man. But, unlike the author of the thriller, you should not leave this obvious fact to be inferred. You should say it.

If you do not say that the factor of safety is 20, John Smith must still think it. I will admit that he is well able to do so. He knows what the breaking strength of the material is and can quickly recall it from the storehouse of his memory. The mental arithmetic gives him no great difficulty. He can work out for himself perfectly well how many times 25 goes into 500. So he can reach the conclusion that the factor of safety is 20 even if you do not tell him so. But he cannot attend properly to the next part of your report at the same time. So I submit that you ought to tell him. To be functional you should add the words 'which gives a factor of safety of 20'.

You might even add the further words 'and this is ample'. But to do so would, I think, be to apply my rule too rigidly. Every rule must be observed with judgment. A good functional style is one in which the writer is explicit enough to be helpful, but not so explicit as to be irritating. That a factor of safety of 20 is high can be appreciated instantly by your John Smith without noticeable effort.

Comment words

There are other occasions when the significance of a number is not so immediately apparent. On such occasions you should say whether the number is high or low or to be expected, even though John Smith may be able to find it out unaided. There are many ways of doing so. Instead of saying 'the temperature is 400°C' it may be better to say 'the temperature is *only* 400°C', or '*as high as* 400°C', or '*unfortunately no more than* 400°C', or '*as one should expect*, 400°C', or '400°C, *a figure from which it would not be wise to draw any conclusions*', or '400°C, *which suggests the need for a change in the method of operation*'.

I have pointed out on a previous occasion that the presentation of numbers calls for particular care because the work that has to be done on them is so much greater than that usually needed for purely qualitative information. The one or two informative little words suggested above may not always be sufficient. It may be wise

Making it easy to understand

on occasion to find more elaborate means of isolating a number effectively from its surroundings, so that it may stand out and attract for a moment John Smith's concentrated attention. The words used to frame the number and isolate it should be so chosen that they do John Smith's thinking for him. They should state the conclusions about the number that you want him to reach; they should explain its significance; compare it with other quantities, if need be; mention any obvious courses of action to which they point.

Though not always so essential, a similar technique is often helpful when non-numerical information is being presented. One additional word, the little word 'even', for instance, may save John Smith quite a lot of trouble and puzzlement. If you say 'lamp black does not provide a perfectly black surface', you may mean that lamp black does not provide a perfectly black surface while other materials do. Or you may mean that lamp black is the best material for the purpose but still imperfect. John Smith has to guess which it is. But if you say '*even* lamp black does not provide a perfectly black surface', he will not be left in doubt on that particular point. There are many similar contexts in which the little word 'even' can do much towards maintaining receptivity.

There is no limit to the number of other words that may serve the same purpose. You can point to the conclusion that is to be drawn from the information by the addition of words like 'fortunately' or 'unfortunately', 'surprisingly' or 'as is to be expected'. Other, more specific, more thoroughly informative, words can be even more helpful. I can imagine a situation in which it would be good to say, 'and yet the points all lie on a straight line', while it would be even better to say, 'and yet the points all lie, *deceptively*, on a straight line'. Similarly it might be good to say, 'the temperature reaches the high figure of 400°C', and better to say, 'the temperature reaches the *dangerously* high figure of 400°C'.

I think that words of this type are important enough to be remembered by a name. So let me call them *comment words*.

It is often helpful if you convey by words whether a statement is a reminder of what John Smith knows already or a piece of new information. If it is known to him, but presented in a form to suggest that it is something new, he is puzzled and irritated. If it is new, but presented in a form to suggest that he ought to know it, he is also puzzled and perhaps also irritated. Quite a simple statement may leave some such disturbing doubt as to its significance. 'This substance is slightly alkaline', for instance. Better alternatives may in certain circumstances be, respectively, 'this material is, *of course*, slightly alkaline' or 'this material, *it should be explained*, is slightly alkaline'.

Such helps to understanding are obvious and given instinctively by many writers. But many do not provide them nearly as often as they might. I have had to pause frequently during my reading to ponder over the significance of some simple statement where one comment word would have saved me the trouble. I must confess, on re-reading what I have myself written I have frequently caught myself committing the same sin of omission. So I think the need in functional writing for plenty of comment words wherever they can be helpful is well worth pointing out.

The fault of excessive objectivity

When they are omitted it is, moreover, not always from carelessness. The fault is committed deliberately, particularly by scientists. It is not that their artistic sensibility makes them shrink from stating the obvious. It is from a misguided devotion to objectivity. The young scientist is taught that he must always remain austerely objective; that he must faithfully set down every observation without prejudice, fear, or favour; that he must avoid the slightest suggestion of personal judgment; that the significance of what he has to present resides in his facts and figures and not in any opinions he may express about them; that, in short, he must always let these facts and figures speak for themselves. Such teaching is part of what is called training in the scientific method.

The young scientist who has taken these lessons about the need for objectivity to heart, schools himself to eliminate from the presentation of his material everything but the bare bones. He would not dare to say that a given temperature was *dangerously* high, or even *high*; for that would be like an expression of personal opinion. He would not say that a given result was either *expected* or *unexpected*; for that would suggest that he had made up his mind about the result before it had appeared. He would not say that a given fact was *worth pointing out*; for that would suggest a partiality for certain facts over

The presentation of technical information

others, whereas in the scientist's world, according to his training, all facts are equally worth pointing out.

I do not want to suggest that such a scientist consciously strips his style of all those comment words that can be so helpful to the understanding. But, nevertheless, excessive worship at the shrine of the goddess Objectivity has impaired his literary style. He avoids the personal so persistently in his scientific world that he cannot admit into that world the personality of any John Smith. He has been warned against the improper use of subjectivity and has learnt the lesson too well. He has taken it to mean that subjectivity is, in science, a deadly sin to be avoided on all occasions. He reminds me of the nonconformist parson whose objection to gambling is so permeating that he will not allow his daughters to play patience.

Bridges of logical reasoning

The one or two little words about which I have been speaking do not, of course, always suffice. Much more may be needed. To ask a great deal from one little word is not to be concise, but to be unfunctional. A good writer attends to the strength of words as a good engineer attends to the strength of tie bars. Single words can do much, but they cannot do the work of a paragraph. There are occasions when nothing less than a 'comment paragraph', or even more, will serve to help make the information easy to understand.

One of the most important occasions when more than a single comment word is needed occurs when a line of

Making it easy to understand

deductive reasoning is being followed, as in mathematics. Then a well constructed bridge of logical reasoning is required. And the little comment word that is sometimes provided to do the work of such a bridge is, of all irritating words: 'obviously'. 'From which x is obviously equal to ... etc.' This is to place a greater load on one word than it can bear. As a bridge of logical reasoning 'obviously' is a failure. Something more complete is needed.

What holds for mathematical reasoning holds for all logical reasoning. Carefully constructed bridges must be provided to every deduction. They are, in fact, even more important when the reasoning is not mathematical. While he is studying any mathematical proof, John Smith is quite prepared to take his time over it. He knows from past experience that he must pause and ponder over and over again. But when he is reading ordinary words he does not expect to have to do so. This is why it is so important that you should guide him, step by step, from the facts to the conclusions. This should be done at the rate at which *his* mind works, not necessarily at the rate at which *yours* does. If you do not do this he will try to bridge the gap unaided. 'Why does it follow?' he will be asking himself. 'What is the connection?' While he is thus preoccupied your discourse passes on and on. John Smith will miss much of it on account of your inconsiderateness.

So, having presented the facts, do not be content merely to continue with: 'From which it follows that...' Perhaps it does follow. Perhaps, even, John Smith can see that it follows. Perhaps he is the type of mental acrobat who can jump across wide chasms in the land of thought and alight exactly where he should, who can reach conclusions intuitively. But do not depend on his mental agility. If you do you will often find that he is left behind, struggling to find a way across the gap, while you have strode on, far out of sight.

A great deal of technical literature would be easier to follow if more bridges were provided between facts and conclusions. So why are they omitted so often?

Sometimes, I think, out of mere carelessness. Sometimes because the author is by temperament a soliloquiser. And sometimes from a sort of self-consciousness. An author may feel that assiduous bridge building should be left to the writers of popular books, that it is a form of pandering to the uninitiated. He may fear that his eminent colleagues would consider themselves to be insulted if he were to make his conclusions appear too easy. But I do not think they would; and I hope that they would rather be insulted than left in ignorance.

The discipline of bridge building

There is, I am sorry to have to admit, sometimes yet a further reason. The bridge is omitted from an author's presentation because it has never been constructed in his mind. We all commit that sin of omission more often than we think. We tend to assume that bridge building is a form of pandering to the slow-witted and that our own swift-winged thoughts can dispense with such pedestrian aids. We should remind ourselves, if we are

Making it easy to understand

wise, that human reason is a fallible instrument and human intuition a still more fallible one.

So one of the hints that I would give to every expositor is this. Be watchful of all your conclusions. Make sure that a bridge leads to each one of them. If there is none, if you have jumped to the conclusion, build a bridge at once. Build it not only in pure thought; that is liable to be too intuitive. Build it in words. Build it so completely that you can reach the conclusion by quite small, closely reasoned steps. Do so even when such a labour seems to you unnecessary, when you feel sure that it is perfectly safe to jump to the conclusion.

If you can bring yourself to do this the result will, I venture to suggest, often prove disconcerting. You will find, to your dismay, that the bridge does not take you to the place that you expected to reach.

Words, if chosen conscientiously, sometimes do for us that unwelcome service. They guide us to the humiliating discovery that what had appeared previously to be obvious is, in fact, not even true. For words are not only a means of expressing thought. They are also a means of creating it, and when they do not create they clarify. They do so by the discipline they impose.

But only, of course, so long as they are treated properly. Treat words badly and they lead you badly astray. Treat them well and they serve you well. So be mindful of the value of submitting to their logic. When an author resents the discipline of precise words, when he prefers vague clouds of intuition to hard facts of logic, when he is so impatient and so sure of himself that he cannot be bothered to seek precise words for every element in his reasoning, then it is not because he is a powerful thinker; it is because he lacks self-criticism.

In science and engineering the discipline of words will be replaced, sooner or later, by the discipline of hard facts. Observation and experiment are available for the checking of conclusions reached impatiently by over-bold intuitive bounds. But the philosopher works in a realm where the hard facts are not forthcoming. In that realm there is only one way of checking bad reasoning, namely by better reasoning. Logic and the critical faculty must do for the philosopher what observation and experiment do, at least partly, for the scientist: keep him on the straight path.

That they often fail to do so is proved by the large number of rival philosophical systems that flourish side by side, each with its little band of adherents and its little band of opponents. Most of them are mutually incompatible, so they cannot all be right. The wrong ones ought to have been destroyed long ago as false scientific theories are always destroyed in time. But the better reasoning that is needed for their destruction has never been strong enough to destroy them. Hard facts would have done the job in a hand's turn. The moral of which is that the discipline of words is easier to dodge than the discipline of facts.

Subsidiary bridges

Sometimes it is wise to construct more than one bridge, particularly when the conclusion is unexpected. The reaction of an intelligent John Smith to a very surprising

Making it easy to understand

statement to which you have provided a rigorous proof often is, *and should be*: 'I know it is true and I don't believe it'. That is how many scientists felt at one time about relativity and the quantum theory, to take two well known examples. Both were reached by perfectly sound bridges of logical reasoning and yet few scientists were quite satisfied with those bridges. It was not that they found any flaws in them, but that they were conscious of the fallibility of human reason and always distrusted single bridges. Today, of course, both relativity and the quantum theory can be reached from a variety of directions, over a number of different bridges. Those who were sceptical at one time are now convinced. This is not because the original bridges of strict mathematical proof have been perfected, but because so many additional ones have been built. Though they are inferior ones, it is these that have carried conviction.

Examples are, of necessity, only to be found in rather recondite and highly specialised fields, so any that I may give cannot be wholly comprehensible to all. But I am hoping that one or two examples, even if somewhat obscure, will illustrate sufficiently what I mean by alternative bridges.

Many of us who found the quantum theory hard to swallow in its early days were greatly helped when someone pointed out that a quantum theory ought to have been expected because of a very simple consideration. If, as had been previously supposed, energy changes were not quantised, but continuous, the whole of the potential energy in the universe must have been converted into kinetic energy in a very short space of time. That it had not so happened suggested forcibly that energy changes could not be continuous. This argument, which I need not develop fully here, substantially increased the receptivity of many of us for the quantum theory.

Waveguides may provide another example. Nowadays electric waves are caused to travel along the inside of rectangular or circular metal pipes as though they were water. But in our youth we were all taught that electricity cannot exist inside a metal conductor. Such a conductor was called a Faraday cage and it was clearly shown by calculation and experiment that the whole of the electric charge was on the outside. Yet, although the waveguide is truly a Faraday cage, the electric waves do not only exist in it but travel along it. I think the reaction of many of us both to the mathematical and the experimental proof tended to be, as suggested just now: 'I know it is true, and I don't believe it'. But I well remember how my receptivity to the theory of waveguides was helped when a friend built me an alternative bridge. He pointed out that the walls of the waveguide are, in effect, a system of quarter-wave lines. From such a system one should expect exactly what is known to happen in a waveguide. Never mind why. The point is that the theory of quarter-wave transmission has long been well known to all electrical engineers. A bridge that uses this theory is constructed of familiar materials.

Let me give one further example. It can be shown mathematically that when an electric transmission line is very long, the current flowing into it at the sending end is the same whether the receiving end is short circuited or open circuited. Students, I have found,

Making it easy to understand

understand the proof quickly enough, for it is not difficult. Yet they remain puzzled by it, particularly the brighter ones. Till then they had thought that the current flowing into the sending end of a transmission line would always indicate quite clearly whether the far end was short or open circuited. They would expect a destructive rush of current to flow into the short circuited line and none into the open circuited one.

This puzzlement causes the student a considerable amount of trouble. A second bridge, additional to the rigid mathematical proof, relieves him of it. One such bridge is provided if one says: 'The charging current in a long line is great enough to swamp the effect of any current due to a short circuit'. Another bridge would be to say: 'There can be no means of discovering the conditions at the far end of an infinitely long line. But if the current that flowed into the sending end depended on conditions at the far end, this current would provide means of knowing what those conditions were. So one should not expect the current at the sending end to be influenced by the receiving end conditions in an infinitely long line. And one should not expect the influence to be very great if the line is very long, even though it is not infinitely long.'

Such alternative bridges of reasoning lack the logical rigidity of a mathematical proof and are, not unnaturally, somewhat in disfavour with meticulous scientists. Considered by themselves they would only satisfy those who are loose and superficial thinkers. So these alternative bridges may appear as a form of pandering to the uninitiated. Perhaps they are. But pandering is, I think,

The presentation of technical information

justified if it helps to convey technical information from mind to mind. The opportunities for such alternative bridges are more numerous than the few examples just given might lead one to suppose.

A few hints

So far I have discussed only in somewhat general terms the methods available for making one's discourse easy to understand. Perhaps it will be useful if I add a few do's and don'ts on the same theme.

DO, during the development of a mathematical argument, occasionally point out the significance of terms and formulae encountered in the course of the argument. True, the final result may alone interest John Smith. But he is, nevertheless, helped and reassured if you tell him something about the route. When opportunity offers, do so with such remarks as: 'This term represents the potential energy in the system and this one the kinetic energy'. 'This expression disappears when the system just fails to oscillate.' 'It will be noticed that at the limit a equals b.' 'In this expression we recognise the Z-factor.' Comments of this nature are more plentiful with good than with bad expositors. By their judicious use you can make your discourse both easier to understand and more fascinating.

DO make plentiful use of graphical methods of presentation. If you are a first-class mathematician you can probably dispense with them yourself and you may then view them with some disfavour. For a mathematical statement conveys a very general truth and a graphical

Making it easy to understand

representation only a particular application of it. So you will rightly feel that by letting curves and diagrams take the place of letter symbols you are narrowing down your statements to special cases. But it is worth while to do so in the interest of good exposition. For graphical methods are of great help to those who cannot think in completely abstract terms.

DON'T, when using curves and diagrams, allow them to contain more detail than is needed for the purpose of your discourse. It is often difficult, especially when preparing drawings of mechanical systems, to put in all that is needed as an aid to the understanding and to leave out all the rest. But it is worth while taking trouble with this difficult problem.

The advice is all the more worth emphasising because it is so fatally easy to ignore it. Suppose you are describing a particular kind of arc-extinguishing device in an oil-filled circuit breaker. A working drawing of the circuit breaker has already been prepared and the device is shown in this. But the drawing was made for a different purpose and shows many other things as well: all the levers and crossarms and toggles and baffles and tanks and other details of that tricky piece of apparatus. The temptation is great to use the available drawing, from which, admittedly, John Smith can extract the necessary information about the arc-extinguishing device if he spends long enough over the task. But remember that what saves you labour gives John Smith much trouble. If you are a good expositor you will make a new drawing in which only the arc-extinguishing device is shown.

DON'T place too many curves on the same figure,

especially when they intersect at acute angles. And don't have too many scales on the coordinate axes.

DO label coordinate axes with the units that they represent and the scale (see Chapter XVI for how this should be done). Do so even if you think that John Smith ought to guess what the axes represent. He might guess wrong. Besides, he wants to receive technical information, not to engage in a guessing game.

DO, also, give informative titles to all illustrations. The title may be clear from the text and this may refer correctly to figure numbers. But John Smith does not want to have to scan the text in order to know what the pictures mean.

DON'T, during a lecture, have a figure thrown on the screen to illustrate what you have just been saying while you proceed to the next section of your lecture. That causes the eyes of your audience to be concerned with one thing while their ears are concerned with another. Eyes and ears can do this but brains cannot. The audience must either take notice of the slide and miss what you are telling them or listen to you and ignore the screen.

DO, when preparing pictures, diagrams, graphs, tables, etc., for lantern slides, remember that the audience will not be seeing them for long. Limit, so far as possible, the information contained in each slide to what you want to convey during the lecture. Do not, for instance, expect the audience to appreciate the significance of all the numbers contained in five columns, or to disentangle a dozen intertwining curves. It is sometimes worth while, when a paper is to be published, to prepare two sets of

Making it easy to understand

illustrations: a simple one for the audience and a more informative one for the readers.

DO, when describing mechanical devices, use all your inventiveness in the search for common objects to which to compare the various bits and pieces that you have occasion to describe. It is usually difficult for John Smith to visualise these from a verbal description; and if their significant features are three-dimensional, even drawings do not solve all his difficulties. You can help him if your description makes plentiful use of terms like 'diamond shaped', 'lozenge shaped', 'saucer shaped', 'in the form of a circular rubber stamp', 'dumb-bell shaped', 'a truncated cone shaped like a lampshade'.

May I add this: if it is important to be inventive of means of expressing shapes for engineering devices, it is even more important for anatomical objects. For in anatomy significant shapes are always three-dimensional and very complicated. Yet, to judge from such anatomical textbooks as have come my way, medical authors spend at best minutes in the search for suitable ways of describing complex shapes when they ought to spend hours on the task. The weary hours have to be spent by the students who would master the textbooks.

DO, when describing a mechanical device or invention, consider very carefully whether first to describe the mechanical details and then to say what purpose they serve or first to state the problem that the device is designed to solve and then to show, while describing the mechanical details, how the purpose is served by them. I do not think that one can make a rule and say that one of these methods is always better than the other. The choice should depend on circumstances. First to enumerate and describe the levers and cams and grub screws and whatever else belongs to the device and then to say what they are for is perhaps the more straightforward procedure. But to state the problem first and then show how it has been solved is the more stimulating one. It gets John Smith trying to do some inventing for himself. It puts him in a state of expectancy, almost as though he were reading a mystery story. When he is told 'the purpose of this device is to do so and so' he begins to wonder how it can possibly be done. This causes him to be very receptive to every word that tells how the problem is solved.

DON'T, lastly, think that an abundance of explicitness is appropriate only to the lower level on which popularisers of science work. It is, admittedly, possible to overdo explanations and thereby become irritating. But over-explicitness is not a common fault and under-explicitness is. Besides, you can afford to be very explicit indeed without irritating if you contrive to be stimulating at the same time.

IX Making it easy to remember

Unretentive audiences

'I was at a most interesting lecture yesterday. Such a delightful speaker. So clear and stimulating.'
'Really? What did he say?'
'Oh, I can't remember what he said.'

I want to be fair. When in previous instances your discourse has not been properly conveyed to John Smith I have always blamed you. But if you are the delightful speaker who provided the occasion for the above dialogue and John (or Mary) Smith was the enchanted member of your audience, I will allow that it is probably not your fault if John or Mary does not remember today what you said yesterday. There are those who make a habit of going to lectures. They obtain much satisfaction from this agreeable pastime. But they can hardly be said to listen to the lectures. They bask in them. The above John or Mary may be among these.

Baskers at lecturers like to have things explained to them. Not that they are interested in the explanations or able to follow them. But it gives them a pleasant sensation when science passes over their heads. They then feel themselves to be in the company of great and important

The presentation of technical information

things. There is something stimulating and consoling to them in the thought of the wonders that have been laid bare. And the more incomprehensible these wonders are to them the better they like it. They derive a particular thrill from astronomical figures. They think it is most instructive when they hear the words 'millions of millions' repeated several times. The information that a cube of 1 cm side contains no less than about 26 million million million molecules would be received by them as yet another proof of the astounding marvels of science.

If some have the habit of *hearing* much and retaining little, others have the habit of *reading* much with no different result. For one can bask in books as well as in lectures. But not so happily, I think. Book basking is too solitary an indulgence. It does not give the same feeling that we are all moving together, striving, climbing to higher and better things. To be a really satisfactory pastime, basking in words must be enjoyed in a crowd.

The circumstances that make information easy to retain

However, I am not concerned here with, or for, baskers of any type. To mention them at all was merely a digression. No amount of trouble on the part of the speaker or writer can change them. And they are not worth troubling about anyway. I am concerned only for the honest John Smith who genuinely wants to profit from what you are able to tell him. To profit he must remember it. We all know, from our own experience when we have been that John Smith, that he remembers it with

Making it easy to remember

varying success; sometimes with ease, sometimes with difficulty, sometimes hardly at all. On what does the difference depend?

Partly, of course, on himself. Lack of effort, lack of interest, lack of understanding may be the causes of forgetting. But these causes are, at least partly, under your control. You can stimulate him to the necessary effort, you can awaken his interest, you can help his understanding. Persuasiveness and clarity go a long way towards making your discourse easy to remember. But I am not sure that they go all the way.

Persuasiveness and clarity certainly help John Smith to *receive* the information that you have to impart; but something additional may be needed to help him to *retain* it. The latter purpose is best served, I think, if one or more of the following conditions is met:

(*a*) if the information is accompanied by associations with which the person addressed is familiar,
(*b*) if it is relevant to the part of the discourse in which it occurs,
(*c*) if it is suitably timed,
(*d*) if it has been carefully prepared,
(*e*) if it has been eagerly awaited,
(*f*) if it is judiciously repeated,
(*g*) if it is memorably phrased.

All these conditions are under the control of the person who is imparting the information. Let each be considered in turn.

Associations

Anything that arouses powerful associations is remembered without difficulty. If you say to John Smith: 'Your house is on fire', or: 'Your uncle has left you a hundred thousand pounds in his will', there is no need to add: 'Now don't forget it'. But if you say to him: 'The purpose of a guard ring is to equalise the stress distribution over a chain of suspension insulators', his effort at memorising the information may be appreciable.

The explanation is simple and has been given at the end of Chapter V. What is remembered is tied, as it were, to existing knowledge. This has to be present in consciousness in the form of associations at the moment when the act of memorising is being performed. If the associations are many and powerful, such as those instantly evoked by mention of a fire or a legacy, the work of memorising is practically effortless. If hardly any associations are evoked, as happens when one hears a sentence in a foreign tongue, the work of memorising is very difficult. It can only be achieved, as mentioned previously, by the creation of those artificial associations that are called mnemonics.

Guard rings and chains of suspension insulators come, for the John Smith who we are now imagining, between these extremes. They are not, it is being supposed, matters of daily concern to him. He can understand what you mean by stress distribution, but the words do not come trippingly from his lips. That is why he has some difficulty in memorising the information that the purpose of a guard ring is to equalise the stress distribution over a

Making it easy to remember

chain of suspension insulators. The associations that come unaided to his mind are too few and too pale.

This leads to one of the rules of presentation. *If you want the person addressed to remember a piece of information make sure that he has sufficient associations to tie it to.* If the information is not likely to evoke a sufficient number by itself you should say things that will evoke further ones. To do so, you say things about the information that are calculated to produce a vivid impression. They do not necessarily help to define your terms of reference or act as aids to understanding. Their sole purpose is usually to act as aids to memory. In our example, a thousand things about guard rings, stress distribution, and suspension insulators might serve this purpose.

In your selection of associations designed to aid the memory it is, moreover, useful to remember a characteristic of human psychology: *It is easier to tie a memory to a picture than to an idea.* John Smith will, for instance, remember what you tell him about uneven stress distribution more easily if you show it in a graph, than if you merely say in words that it follows an exponential law. And it may help him to remember what you say about guard rings if you do not only speak of them in terms of ideas but also remind him that he has seen them at the bottom of chains of suspension insulators on his country walks.

Authors of imaginative literature exploit very cunningly this peculiarity of the human mind. Their writing abounds in visual descriptions. Do not copy them slavishly. Do not mention the appearance of things as often as they do. But do sometimes use descriptive adjectives like tall, red, shiny, curved. Expand them now and then to a descriptive sentence or two. If it can help John Smith's memory after a brief description of suspension insulators to add: 'They are usually green', do not hesitate to do so. Do not allow self-consciousness and an exaggerated idea of the need for austerity in scientific presentation to bar the occasional use of an evocative turn of phrase. Bear in mind that it is your duty as expositor to help the memory as well as the understanding. But do not overdo the evocation of visual associations. If you overstimulate John Smith's visual imagination you may divert his attention from the subject matter.

Relevance

Remember this: a digression, a parenthesis, an aside will be forgotten, even if it is interesting, when quite a dull remark in the main current of your discourse will be retained. The reason is, here again, in the word 'associations'. John Smith has assembled in the conscious compartment of his mind those associations that are required for the main current of your discourse. Anything that is relevant to this main current can be tied at once to the associations that are there ready for it. But nothing is available to tie irrelevant associations to. So the storekeeper must hurriedly bring the required ones along only to pack them away again a moment later. If he can manage it once or twice he cannot do so repeatedly. Nothing destroys receptivity more than the constant packing and unpacking of associations.

Making it easy to remember

This is the chief reason why it is important to construct one's discourse on logical lines. If it is so constructed the right associations are always available to help the retention of each item of information. If it is not so constructed many items will be soon forgotten. Perhaps they will not even be noticed. Here the rule of presentation is: *Select with great care for everything that you have to say the proper place for saying it.* The proper place occurs at the moment when John Smith is thinking about the theme to which the thing you want to say belongs. Hence the importance of architecting, to which I have referred already towards the end of Chapter VII.

Timing

Your discourse can only be remembered with ease if you show consideration for the metaphorical storekeeper. He has to wrap up parcels of newly acquired information and put them away on their proper shelves. This, like all mental activity, is not accomplished without some effort; and there is a proper time for it.

The proper time is not while John Smith is otherwise occupied. It is not, for instance, while he is learning what your terms of reference are; for then the storekeeper is too busy collecting associations. Nor is the proper time for memorising while John Smith is trying hard to understand the meaning of what you are telling him. While he is concentrating on this work he has no effort to spare for taking mental notes. In dirty weather the man at the wheel is not prepared to enter into small talk with a passenger. And John Smith struggling with new information is like the man at the wheel. Only when he sees the ship of your discourse glide past the beacon that marks the entrance to harbour and still water, does he find the leisure for mental note taking.

So it is a part of an expositor's duty to provide moments of harbour and still water. The things said during these moments must be designed for the one and only purpose of aiding the memory. They must not require a substantial change in associations and they must not place a substantial strain on the understanding. They must be repetitions of what has been said already. But, as I shall explain in a moment, not verbatim repetitions. They must be what I propose to call judicious repetitions.

An obvious occasion for them is at the very end of your discourse; one expects to find harbour and still water at the end of a voyage. This is why many reports and papers conclude with a grand summary. In this the most noteworthy statements are repeated in quick succession.

Such a summary is often useful; but it is not enough. John Smith must be given an opportunity for memorising each item as and when he meets it, while the associations to which it is to be tied are still assembled in the conscious compartment of his mind. So you should ask yourself after setting down each item: Will John Smith memorise this instantly or ought I to provide for him a moment or two of harbour and still water? You will find that you should do so quite often at the end of a paragraph and sometimes even within the body of a paragraph.

Making it easy to remember

Preparation

By preparation I mean the use of a form of words to indicate that the remark to follow is designed for the sole purpose of aiding the memory. There are many such forms of words. Among the more hackneyed are: 'To sum up', 'in fact', 'thus', 'in short', 'let me repeat', 'in other words'. Such expressions are, in the metaphor used above, *beacons* that mark the entrance to harbour and still water. When John Smith encounters one he knows that during the next moments he will neither have to collect new associations nor try to understand something. He will only have to memorise something. He makes his mental dispositions to suit.

For instance, 'the stress distribution follows an exponential law' is, let us assume, but a recapitulation of what has been explained just before at some length. But cast in that form it may not be immediately recognised by John Smith as a mere recapitulation. He may think for a moment that it is bringing something new for him to worry over. You will save him from this uncertainty if you provide a beacon with the words 'in fact'. 'In fact, the stress distribution follows an exponential law.' Words used in that way can be a valuable aid to the memory and as a technical term for them will be useful, I shall call them beacon words.

A skilful author invents other more ingenious, more subtle, beacons. The form they take is one of the characteristics of his style: sometimes a very unobtrusive sign that harbour is near suffices. A colon may be such a sign, or in rare instances only a comma.

Anything helps that indicates when a part of the journey is approaching its end and harbour may be expected. The sight of the end of a paragraph does so, for instance. So may the sight of a short pithy sentence after a long elaborate one. As it has the form of a recapitulation John Smith assumes that it is one. There is often no need to tell him so.

Expectancy

A piece of information is remembered more easily if it has been eagerly awaited than if it has slipped in unannounced. So one way of helping John Smith's retentivity is to create in him a condition of expectancy. Before you present him with the information you arouse his curiosity. Thereby you cause him to collect associations in advance, to which the information will be tied.

To do this you use a form of words to show that you are leading up to an important statement. You have just explained, let it be supposed, that the stress distribution over a chain of suspension insulators follows an exponential law. You show next that the consequence of this law may be a flashover that begins with the bottom insulator. Then you ask a question: 'What can be done to overcome this unfortunate circumstance?' John Smith is caused thereby to want to know the answer. When told that it lies in the provision of a guard ring, he associates the information evermore with the curiosity with which he awaited it. Thereby it becomes tied more firmly to his memory.

Making it easy to remember

Judicious repetition

Sometimes an item of information can safely be given once only. But when this is done the acts of understanding and memorising have to be almost simultaneous; and this may impose too great a strain. So it is often better to say the same thing more than once. When this is done the second statement occurs in what I have called above a moment of harbour and still water, after the hard work of understanding has been completed, and constitutes what I shall call a judicious repetition.

I say judicious advisedly. It may be true that mere verbatim repetition of a statement makes it easier to remember, just as a nail holds more securely after the second hammer blow. But I do not like the analogy. The methods that you employ to make your discourse memorable should be analogous to some more delicate instrument than a hammer. You do not want to drive the information into John Smith's mind with hammer blows. That will eventually only dull his wits. You want to stimulate his mind to the activity that constitutes memorising. You only do this when you avoid verbatim repetitions and bring instead something new at each repetition of the information.

Something new, but not much. There must be just enough to maintain interest and not enough to impose any substantial strain. A moderate amount of novelty causes a few further associations to be brought into relation with the information. These serve to fasten it more securely to the shelves of the memory.

The most superficial kind of novelty that may serve is

The presentation of technical information

purely visual. This may, for instance, be provided by the appearance of the page on which a grand summary of your discourse is set out. John Smith will associate the items summarised with the picture presented by the headings, the typescript, the margins, the spacing. For this reason alone it is important that a summary page should be pleasant to look at. It should present a pattern to dwell in the mind. The heading should not be too near the top of the page; it should be carefully placed centrally; the typing should not be too crowded; numbered items should be clearly distinguishable; capital and small letters should be appropriately selected; the layout should be symmetrical. If such matters are properly attended to John Smith will be helped to remember particular items of information by associating them with their positions on the page.

If in a summary items are numbered it may also serve as a useful, though superficial, aid to the memory. If you say: 'To sum up, four uses of this appliance can be distinguished', John Smith will add to his previous associations a number for each of the uses. It does not seem much to add, but it is a common experience that an item springs into recollection in association with a number when it fails to do so in association with an idea.

However, the grand summary is not the only form of judicious repetition. Such repetitions should occur, as I have mentioned already, many times during your discourse; in fact at all those moments that I have described as harbour and still water.

The repetition may consist in a mere change from a positive to a negative form of words or *vice versa*. 'The

Making it easy to remember

points do not all lie on a straight line. There is a distinct curve.' The method can be justified occasionally; but it is dull. It provides repetition without sufficient stimulation. So there are occasions when it would be better if the second sentence were: 'There is a distinct curve upwards'. 'Upwards' stimulates a visual association that the word 'curve' alone does not stimulate. Similarly it might be just serviceable to say: 'This substance is not colloidal. It is crystalline'. While it would be better to say: 'It is crystalline, like quartz'.

Alternatively a classification may serve to impress facts on the memory: 'To sum up, these insects have such-and-such characteristics. They must clearly be classified among the orthoptera'. Or a comment may serve the same purpose: 'Such, then, are the facts. Unfortunate, but unavoidable'. There is no limit to the variety of mildly stimulating additions to an item of information that may serve as mnemonics.

Sometimes a striking analogy is both an aid to the understanding and an aid to the memory. But it serves the latter purpose only, of course, if it is presented in harbour and still water. In such a position the analogy repeats in another form what has already been presented in direct statement. Suppose, for instance, that you have been developing a formula that defines the performance of a certain electrical circuit. You complete your explanation with the words, 'It will be seen from the final equation that the circuit fails to oscillate when the leakance exceeds a certain critical value'. This is the fact that you want John Smith to remember. You might aid his memory by a judicious repetition in the form of this analogy: 'When the leakance is above this value the circuit could be described as "soft". It is then analogous to an old tennis ball that fails to bounce'. I can picture a John Smith who would be helped by such an analogy not only to understand better, but also to remember more surely, the significance of the critical value of the leakance.

Here is another conceivable example. You have occasion to say: 'Radiation of wavelength 12.5 mm is absorbed by water'. With those words you would have said all that you needed to in order to convey the information. But yet you might, I think, aid the memory if you were to continue with the slightly stimulating words 'as blue light is absorbed by a sheet of red glass'. Here the idea of optical absorption, which is what you want to convey, is repeated judiciously in a form that brings it into association with the continuity of the spectrum over the entire range of wavelengths.

What in imaginative writing one sometimes finds in the position that I have described as harbour and still water is an illustrative example. From this one might think that it serves as a means of memorising the statement that it illustrates. But I doubt if the illustrative example ever serves this purpose. True, it repeats the statement in another form. But this is not a form that aids the memory. Let an example explain my meaning. The paragraph from Bernard Shaw of which I quoted the first sentences on page 80 will serve:

'Wellington said that an army moves on its belly. So does the London theatre. Before a man acts he must eat. Before he performs plays he must pay rent.

Making it easy to remember

> In London we have no theatres for the welfare of the people; they are all for the sole purpose of producing the utmost obtainable rent for the proprietor. If the twin flats and the twin beds produce a guinea more than Shakespeare, out goes Shakespeare, and in come the twin flats and the twin beds. If the brainless bevy of pretty girls and the funny man outbid Mozart, out goes Mozart.'

The significant statement in this entertaining paragraph is: 'In London we have no theatres for the welfare of the people; they are all for the sole purpose of producing the utmost obtainable rent for the proprietor'. The first two sentences are, as I have explained already, theme sentences; they say what it is about, namely economic considerations. The next two sentences are repetitions of the theme sentences in the form of illustrative examples: 'Before a man acts he must eat. Before he performs plays he must pay rent'. They do a good deal to aid the understanding; but I doubt if they do much to aid the memory.

The same holds, I think, for the last two sentences in the paragraph. At least the first of them helps to make Shaw's meaning clear. It might have been missed if he had not added to the significant statement: 'If the twin flats and the twin beds produce a guinea more than Shakespeare, out goes Shakespeare, and in come the twin flats and the twin beds'. This sentence is a functional one in so far as it says 'what it means', but hardly serves the function of 'making it easy to remember'. The last sentence of all is a repetition of the same idea in the form of a second illustrative example. It is not needed as an aid to the understanding and does nothing to help the memory.

So the repetitions in the above quotation are not functional; they belong to imaginative literature. Like the repetitions used so effectively in the psalms, they evoke an atmosphere, create a general impression, induce a mood. When one attempts to recall what Shaw had to say in that paragraph one must use one's own words and not his: 'In the London theatre artistic considerations have to yield to economic ones'. Had Shaw's aim been to write functionally he would have said this himself in the last sentence instead of what he did say about the bevy of pretty girls, the funny man, and Mozart. But then Shaw's aim was not to write functionally: he was, after all, not presenting technical information.

Memorable phrasing

The best form of judicious repetition in functional writing is, without any doubt, one that I shall describe as memorable phrasing. It is also used in imaginative writing, of course. There an author has innumerable devices at his disposal for making a statement memorable. There are the picturesque turn of phrase, the apt analogy, the epigram, the surprising twist, the use of an unexpected word, the anticlimax, the literary or biblical allusion, the misquotation of some well-known saying. In the presentation of technical information such devices are usually out of place. But only for one reason. They are too potent. They stimulate the imagination too vigor-

Making it easy to remember

ously and deflect attention thereby from the information that is being presented. In consequence they irritate; they are disturbing. In metaphorical language they may be said to steal the limelight. A noteworthy scientific statement is most memorable if presented in unobtrusive wording.

When casting a statement in terms that can be described as memorable phrasing the rule to follow is this: *Always set down any information that you want the person addressed to remember in the words in which you want him to remember it.*

If *you* do not do this John Smith must do it for himself. To benefit from your discourse he must muse on these lines: 'Let me see. What was the important point that the author made so clearly? I must try to get the gist of it. Yes. I think I remember now. I should put it this way.' 'This way' is a memorable phrasing, a concise statement, that John Smith has, after some effort, managed to invent for the purpose of memorising an item of information that you have presented. You ought to have spared him this effort. *You* ought to have given him the gist of it. He ought to be able to remember it in *your* words and not to have to invent words of his own.

A piece of information cannot always be presented in memorable phrasing when it occurs for the first time. Then it has to be encumbered with the words needed to explain its significance. This is why memorable phrasing is usually a form of judicious repetition to be presented in harbour and still water. Great care must be taken with its construction so that it may properly serve its purpose. The form must be terse and cogent; there must be no redundant words; the syntax must be crystal clear;

every noun, adjective, and verb must be carefully selected for the exact meaning that it conveys; prepositions must be used with meticulous attention to their logical implications, and not loosely as though one were as good as another. Lastly, the sound of the words and their rhythm must be pleasing to the ear; for even the least sensitive among us is better able to commit pleasant, rather than unpleasant, sounds to memory.

I hope I shall not be considered too flippant if, as a familiar example of a repetition cast in memorable form, I quote the famous 'predestination' limerick. It illustrates neatly several of the points made in the preceding pages.

> *There was a young man who said Damn!*
> *At last I've found out that I am*
> *A creature that moves*
> *In determinate grooves,*
> *In fact not a bus but a tram.*

The significant statement in this limerick is that man is a creature that moves in determinate grooves. The last line forms a judicious repetition of the statement. Three characteristics show that this last line is presented in harbour and still water. There is the fact that it is the last line, a natural place for journey's end and a harbour. There is a beacon 'in fact'. There is the epigrammatic form of words. The repetition is cast in the form of a striking analogy, a form designed to stimulate the act of memorising. The wording of this line has all the qualities of terseness, cogency, and euphony that render it easy to remember.

X Circumlocutions

Everyone who wishes to acquire the discipline of functional English should read the late Sir Arthur Quiller-Couch's famous and most entertaining essay on 'Jargon'. In it he castigates circumlocutions and selects, in particular, as a target for his wit and good sense one of the commonest of them: 'in the case of'. There are, of course, many others: 'from the point of view of', 'in regard to', 'in this direction', 'with reference to', 'due to the fact that', 'in this respect', 'in so far as is concerned'.

None of these cumbersome turns of phrase is entirely meaningless; and each can be justified on occasion. But the occasions are rare. When a circumlocution is used it is nearly always, not because it provides the best way of conveying information, but because it enables an author to dodge thought. So think again whenever one of these phrases slips from your pen. You will find nine times out of ten that you have been lacking in the discipline that an author should always impose on himself.

Having made a little collection of these ugly labour-saving devices, I have reached the conclusion that they may serve the lazy author who uses them in one of four ways. A circumlocution may enable him

The presentation of technical information

(*a*) to dodge the task of saying in a neater way what the information is about,

(*b*) to dodge the search for the correct noun,

(*c*) to dodge the search for the correct preposition,

(*d*) to dodge the task of recasting a sentence that he has begun badly.

Let me illustrate each of these reprehensible practices with the help of an example or two.

Saying clumsily what the information is about

'*As far as* the administration of the country *is concerned*, it never cost the British taxpayer anything.' Here the circumlocution could be left out without making the least difference to the meaning of the sentence. True, the redundant words do help a little to isolate 'administration' from the rest of the sentence and thus tell John Smith what it is about. But the method is clumsy and inadequate. A better means of guiding the storekeeper's steps should have been found.

Dodging the search for the correct noun

'I suspect that he is not aware of what has been done *in this direction*.' The last three words mean nothing whatsoever. The sentence would be better if they were left out. So why did the author put them in? He must have felt, though probably only preconsciously, that he ought to tell John Smith something more. But he was too indolent to think out what it should be. My guess is that it was not clear enough from the context what the situation was about which something had been done. The author ought to have said: 'I suspect that he is not aware of what has been done about the so-and-so'. But the words in which to define the situation, the words for which 'so-and-so' stands in that sentence, could not be found without some trouble. So the glib 'in this direction' took their place.

'The new system would help *in this direction*', is an exactly similar example. So is this one: 'It is desirable for our firm to set an example *from this point of view*'. When you inadvertently write such a sentence your first duty is, of course, to strike the circumlocution right out. But do not let the matter rest there. Try to discover what in your preconscious thoughts caused you to put those meaningless words in. You may find that something less meaningless is really needed, something to help the storekeeper to collect the necessary associations, something that can only be conveyed by a precise and informative noun. Look for it.

Dodging the search for the correct preposition

A preposition defines the relationship between two subjects with some precision. Its correct use demands an appreciation of this relationship, a thing not achieved without a little effort at logical thinking. And a careless or lazy writer prefers not to make the effort. He chooses instead a form of words that can substitute *any* preposition.

'*So far as* Europe *is concerned* Bulgaria and Rumania remain one of the most troublesome problems.' Here the five words of the circumlocution are a substitute for the one word 'in'.

The presentation of technical information

'Under the directives of both Governments the administration of Sicily is to be benevolent *as far as* the civilian population *is concerned*.' Here the same five words are a substitute for 'to'.

'*In so far as* the poultry *is concerned* a drinking fountain is to be provided.' Here again the preposition is 'for'.

'It is interesting to note that comment on the need for training of the character considered in this paper has, *as far as* the Institution of Electrical Engineers *is concerned*, often been incorporated in the Addresses of Presidents and Chairmen of Centres.' Here the correct preposition is 'in'. However, to replace the ugly 'in so far as is concerned' by the one word 'in' would hardly suffice to turn that appalling sentence into functional English. It ought to be recast entirely to something like this: 'It is interesting to note that in the Institution of Electrical Engineers the need for character training, which is being considered in this paper, has often been mentioned by Presidents and Chairmen of Centres in their Addresses'.

Dodging the task of recasting a sentence

The author of the last example had another blameworthy reason for the circumlocution. He had proceeded some way with his sentence when it occurred to him that the Institution of Electrical Engineers had to be mentioned. 'In so far as is concerned' saved him from the trouble of recasting the sentence.

'In this country power stations are now familiar objects to most people *from the* external *point of view*.' Had the author formed the sentence in his mind before

Circumlocutions

he set it down he would hardly have written that. What happened was that he remembered, after he had spoken of power stations as familiar objects, that his statement needed a qualification. It is true only of the external appearance. Most people have no notion what a power station looks like inside. So he tacked on 'from the external point of view'. Had he taken his duties as expositor seriously he would have struck out what he had begun to write and substituted: 'In this country most people are familiar with the external appearance of power stations'.

'We are sometimes inclined to forget that one Distinguished Authority who merely quotes or re-echoes another one, does not make two Distinguished Authorities, *so far as* an original error *is concerned*.' The operative word is 'error'. What the author wanted to say is that an error made by a Distinguished Authority does not become a truth when it is quoted subsequently by another Distinguished Authority. The remark is mildly witty, and I feel sure that its author was very pleased with himself for having thought of it. But such entertainment value as it may possess is lost through bad presentation. The operative word 'error' is delayed until very nearly the end of the sentence. One has the impression that it only just managed to catch the bus. The author was too busy patting himself on the back for his wit to think about his technique. So he set down 'Distinguished Authority' near the beginning of his sentence and twice over, delighted at the bright idea of making his debunking more effective by the use of capital letters for the two words. Then only did he notice that he had not yet described what one Distinguished Authority quotes from another. So he dragged 'error' in forcibly with the help of 'in so far as is concerned'.

'With time, the attackers will be able to improve and stabilise their position, *so far* as space *is concerned.*' I do not think that the author meant to imply that the attackers would *not* be able to improve and stabilise their position so far as anything else was concerned. I think he first set down 'In time the attackers will be able to improve and stabilise their position'. Then the sentence appeared to him too obvious a platitude. So he added a few qualifying words. These might make it appear that he had carefully thought out the various ways in which the attackers might be able to improve and stabilise their position and had finally reached the well-considered conclusion that they would only be able to do so 'so far as space is concerned'. Had the author composed the sentence in his mind before setting pen to paper he would, I think, have given it the form: 'If the attackers have enough time they will be able to conquer enough space to make their foothold secure'. Then it might have occurred to him that he had not much to say.

Enough of these unpleasant things, circumlocutions. We all read examples as bad or worse a dozen times a day. And we are all often guilty of them in our own work. What is truly unfortunate is that we hardly ever notice them. Mankind has evolved an insensitive hide to such ugly turns of speech. We are all so well conditioned that we do not even wince when we meet one. So my main advice is to remind you of my opening remarks on this subject. Study Quiller-Couch and acquire a thin skin.

XI Generalisations

General statements must be supported by examples

The presentation of a broad generalisation raises one or two problems. For a thoughtful person does not accept a general statement until he has tested it. And he does so with the help of particular examples. Without one or more such examples one can hardly appreciate the significance of a generalisation and one can certainly not be sure of its truth. Hence a person who is told a general truth immediately sets his mind to work in search of illustrations with which to verify or refute it.

One need only mention a few generalisations to demonstrate this mental process in the recipient: 'Brightly coloured birds are rarely song-birds'. Who can hear this without calling sundry singing and songless birds to mind? 'Materials expand when heated.' One thinks of materials that do so and possibly of the exception provided by water just above freezing point. 'All insects are terrestrial creatures.' Examples are called to mind at once.

These acts of calling specific examples to mind and testing a generalisation with their help requires some mental work. The work is usually the greater, the more general and the more profound the statement is. This is

The presentation of technical information

why authors who, like Shakespeare, are prolific in generalisations are regarded as highbrow.

If you wish to be a good expositor you must make the work as light as possible for John Smith. One way of doing so is to provide the illustrative examples yourself instead of leaving it entirely to him. So one of the rules of presentation is that one should not make a very general statement without at least one illustrative example. Often several are much better than one.

It is worth while to take much care with the choice of the examples. To be a good one an illustrative example should, I think, have the following characteristics:

(*a*) It should be a true example and not, for instance, an analogy or a metaphor.

(*b*) It should be as free as possible from the complication of extraneous associations.

(*c*) It should be thoroughly familiar to the person addressed.

True and false examples

The first of these three characteristics is likely always to be found in the examples given by a physicist or an engineer when he is illustrating a generalisation within his own field. When presenting the Newtonian principle of inertia he would not refer to the smoker's disinclination to give up tobacco. But I am not at all sure that laymen would not do so. For most of them are mightily confused between the scientific and the metaphorical or colloquial meanings of mechanical terms such as inertia,

Generalisations

power, energy, force. When a physicist or an engineer is outside his own field he may be similarly confused.

I have known both the evolution of the bicycle and the evolution of the solar system quoted as illustrative examples of Darwinian evolution by people who certainly ought to know better. In a scientific periodical I have even read the astounding statement that: 'The process of organic development flows uninterruptedly from atom to man'. Here the change from atoms to simple molecules and from simple molecules to complex ones is treated as an example of an evolutionary change. Yet solar systems do not have ancestors and offspring. No one has seen a bicycle with young. Atoms do not possess reproductive powers. They are not an extinct species of which molecules are the lineal descendants. That the semi-scientific can seriously put forward such false examples, *and not be challenged for them*, suggests that low standards of logical thinking are tolerated nowadays and should be a warning to all of us. Perhaps every one of us is guilty of a similar howler on occasions when he is discussing a subject somewhat remote from his own.

The second characteristic is also likely to be found when a physicist or an engineer is illustrating a general statement in his own field. He would instinctively prefer to illustrate the law of gravitation by the example of a falling stone than a pendulum. For the latter brings in the extraneous association of inertia. It is hardly necessary to say much about this characteristic.

The presentation of technical information

Examples should be familiar ones

The third characteristic deserves a few words of discussion. A scientist, immersed in his studies, thinks naturally first of examples that have formed a recent preoccupation with him. These are likely to be very recondite and not at all familiar to the John Smith whom he is addressing. Scientists who have made a profound study of animal behaviour have, for instance, discovered the following rather important general fact: 'An animal's reflex behaviour is not entirely innate. It is partly conditioned by its experiences'. Those who try to tell laymen what this generalisation means rightly provide illustrative examples. Their favourite one is, I have found: 'Thus it has been proved by science that a dog conditioned to associate food with the sight of an elliptical disc of which the ratio of the two diameters is greater than a limiting number will secrete saliva at the sight of such a disc.' I think they would be better expositors (and clearer thinkers too) if they gave this example instead: 'Thus everyone knows that a dog wags its tail at the mere mention of the word dinner if it is in the habit of hearing that word before receiving food.'

The rule here is this: *When choosing illustrative examples the important point to consider is, what is familiar to John Smith, not what is familiar to yourself.*

Relative positions of general statement and examples

A question that often calls for some thought is whether

Generalisations

the illustrative examples should come before the general statement or after it. Should one proceed from the particular to the general or from the general to the particular? Frankly I do not know. But perhaps one can obtain a useful hint from Shakespeare. His most quoted lines are broad general statements. And they mostly occur as a summing up, as a general conclusion that can be drawn from some particular remark or conclusion.

'Uneasy lies the head that wears a crown' occurs at the end of a monologue in which the truth of that remark is illustrated by the unhappy plight of King Henry IV. Again 'conscience does make cowards of us all' is spoken *after* the example that illustrates it, namely 'the dread of something after death'.

'The apparel oft proclaims the man' is said *after* 'costly thy habit as thy purse can buy, but not expressed in fancy; rich not gaudy'. The general statement that the apparel oft proclaims the man arises out of a specific piece of advice to a particular young man.

In all the above instances the general statement is made in, what I have called in Chapter IX, harbour and still water. Its form is that of memorable phrasing. Its effectiveness depends on a position after the work of associating and understanding has been done. How ineffective each of the above quotations would have been if it had formed an opening statement.

Yet Shakespeare does not *always* place a general statement after the examples. 'All the world's a stage' is an opening remark. It forms what in Chapter VII has been called a theme sentence. The famous string of examples follows. And how trite and ineffective the statement 'all the world's a stage' would be if it had been placed at the end of Jaques' speech. It is not the sort of thing that one wants to be told in harbour and still water. As a summing up it just won't do.

If one can use these cases to devise a rule it is, I think, this: when the generalisation contains a truth that can only be grasped after taking thought, it is better to prepare the person addressed for it. Examples are then a good preparation and they should come first. 'Uneasy lies the head that wears a crown' is a remark for the thoughtful. The thoughtless would expect royalty to be bedded softly and to sleep well. So the remark is more telling when some examples have created receptivity for it. It is unsuitable for the place in which one finds theme sentences; and this for the simple reason that, in popular terms, it is unfamiliar.

When, on the other hand, the statement can be accepted at once without effort, it best precedes the example. 'All the world's a stage' is no great truth. It is an entertaining bit of fooling, not meant to be taken seriously. The trivial scrap of meaning that it contains can be grasped by the thoughtless as soon as they hear it and will be rejected by the thoughtful as soon as they think much about it. The Forest of Arden is no more a place for thoughtfulness than a cabaret. Jaques' speech about the seven ages of man whiles away the time for the ducal audience as a cabaret turn might do. If Jaques were to present 'all the world's a stage' as the final conclusion of his long speech the effect would be rather that of the mountain that laboured and brought forth a mouse. The same rule applies to anything in the nature of a platitude.

Generalisations

It is out of place when employed in any higher service than that of a theme sentence.

However, it would not be safe to derive rules for the presentation of technical information from observations on imaginative literature. Shakespeare did not intend his general statements to become copy-book quotations. He put them in because they were the sort of comment on a situation that his character would naturally utter at that moment. All that I feel sure about is that the right choice is as important in functional as it is in imaginative writing. I am reasonably confident that you will make the right choice whenever you give thought to the matter. Here, as so often in presentation, errors are committed more often because authors do not realise that there are any problems to think about than because they cannot solve the problems for themselves once they have recognised them.

XII On meaning what you say

It happens to all of us from time to time that what we mean to say and what, in fact, we do say are two different things. Often we notice the error in time, as soon as we read over what we have just written. But often we still fail to notice it. Perusing our own words, we read into them what we intended them to mean and not what they actually mean.

This is more liable to happen immediately after we have set the words down, while the intention is still fresh in our minds, than after the lapse of a little time. So it is a good plan to postpone the perusal of one's manuscript until it has faded somewhat from one's memory. Then one appreciates it as though it were written by someone else and one is startled to find here and there a statement that is clearly incorrect. Much of what is written and published nowadays makes the impression that it has never been revised by its author. Probably it has been. But the revision was undertaken too soon and failed for this reason only to be effective.

Failure to say what you mean may arise from sheer carelessness. We often use the wrong word, or perhaps a construction that is ambiguous, if not downright misleading. The failure may arise from a common human

On meaning what you say

weakness, a tendency to exaggerate. It may also arise from another human weakness, a tendency to replace the facts as they are by facts as one would like them to be. There are, no doubt, other causes of the failure. But I think that it is particularly important that you should be on the look out for these three. Let me label them respectively: Careless statement, over-statement, wishful statement. Here are a few examples.

Careless statement

'The continual movement gradually pushes the coal across the grate until, being burnt, the residual ash falls into the hopper.' In this sentence it is stated that the ash is burnt. Of course the author did not mean to say so. The error is one of grammar and rather a crude one at that. It should be avoided by anyone who had reached matriculation standard in English. Many common errors are more subtle and need discussion at a higher level.

'A separator is placed in a convenient position at the outlet from the steam heater in order to ensure that no water enters the turbine, thus causing serious damage.' The author means to say that the entry of water would cause serious damage. But the effect of the sentence is to suggest that the separator is placed in such a position as to ensure that serious damage will be caused. This is typical of a most common error. A word or clause is placed at the end of a sentence and meant to refer to something occurring earlier in the sentence. But the construction is so careless that this word or clause refers in grammar to something else. Here is another example of the same error: 'Whenever the subject is discussed by learned bodies a number of scientists of standing reveal a surprising lack of agreement and ignorance about it.' The author meant to say that the scientists exhibit ignorance. But what he did say was that they exhibit lack of ignorance.

Sometimes an expression is inadvertently used that has more than one meaning while the context is such that the wrong meaning is as likely to be taken as the right one. 'A virus consists of a coating of protein containing the hereditary material.' The word 'containing' should, in this context, mean 'enclosing', but it is much more likely that the reader will think that the hereditary material is actually part of the protein.

Over-statement

Over-statement, to come to the next item, is an error so fatally easy to commit that it requires constant watchfulness. So examine very carefully every superlative term that may have slipped from your pen. When you find yourself saying that something is enormous, consider whether you mean to call it enormous or only rather large. When you say that a certain policy would be disastrous think again. Perhaps you mean only that it is not desirable. When you say that something is essential you may only mean that it would be advantageous. Do not describe all bright things as dazzling, all large ones as infinite, all difficult ones as impossible. Remember that scientists are expected to think and express themselves quantitatively and that words have some quantitative significance as well as numbers. Be economical even with the use of the little word 'very'. Do not say that a thing is 'very big' if you mean only that it is big.

Wishful statement

This sometimes makes unwarranted use of the word 'only'. 'The only way of restoring the national economy is by socialism.' 'The only way of restoring the national economy is by private enterprise.' In politics wishful statements are the rule rather than the exception. So when you detect yourself saying that a given course is the only one to achieve a certain purpose think again. Perhaps you mean the best one, or the only one that you personally like, or the only one that you have so far taken the trouble to think about. For there are probably half-a-dozen ways of achieving the same purpose. Such use of 'only' is a sign of a one-track mind. So remember that the advantages of your favoured way of achieving the purpose will be more readily accepted by the John Smith whom you are addressing if you show him that you have carefully considered and compared all the possible ways of doing it. You say 'only' because you like your own way so much that you wish it were the only one. This is why I classify the wrong use of 'only' among the wishful statements.

Another type of mis-statement that I would classify among the wishful ones makes use of the word 'essential'. You are trying to make out a case for your own point of view. You quote some fact in support of it. And to make the fact seem more important and convincing you describe it as *essential*. Suppose you want to prove that the

The presentation of technical information

earth is flat. You say, perhaps: 'The essential feature of the sea is that it is lower than the land'. You want to prove a theory concerning the age of the earth and you say: 'The essential characteristic of the sun is that it radiates energy at such and such a rate'. I call this misuse of the word 'essential' wishful because it makes, or rather appears to make, your case stronger than it really is.

Now let me quote one last example of wishful statement. It is not an invention of mine, though I shall refrain from naming the eminent biologist-philosopher who is its author. To an engineer or a physicist it may seem to be too extreme to be typical. But I wonder. I fear that those of us who are engineers or physicists may go as far astray as a biologist may when he is discoursing on physics. Here is the passage: 'The very name and conception of matter today gives way to that of energy, doing.'

The context of this statement shows that its author had at heart the theory that physics is progressing from a cold materialism to something higher, something the bishops could approve of, something concerned with *doing* as well as with *being*, something that can be expressed in terms of achievement if not even of values. This theory expresses a wish shared, for some odd reason, by many amateur philosophers in these days.

So the author says that the name and concept of matter today gives way to that of energy. And to make his wish fulfilment more certain he says 'the *very* name'. If it were true, physicists would no longer use the name 'matter'. But of course they do. And they use the concept too. They not only still *talk* about matter, they also still *think* about it.

On meaning what you say

Stripped of its wishful mis-statement, the sentence should say only that matter (or more correctly, mass) is now known to be convertible into energy, just as steam is convertible into water. Because steam is so convertible no one could say that the very name and concept of steam had given way to that of water.

A second wishful mis-statement occurs at the very end of the sentence, where it is declared that energy is synonymous with doing. The wish is to think that energy, which according to this author is the only thing left for physicists to study, is something rather fine, something that one can equate with doing.

Why not 'undoing' instead? For that word would come more readily to the mind of anyone who thought of the energy in a high explosive. However, neither word is, of course, remotely synonymous with energy. In the cold light of hard facts the vague implication of values in our quotation fades away. Any wish to find in modern physics a movement towards higher things is frustrated in the true, but more prosaic statement, that matter is sometimes convertible into a quantity that can be expressed as the product of force and distance.

Make no mistake about it. You and I, every one of us, is guilty of statements of this character now and then. However civilised, however sophisticated, however scientific we may be, we all have, deep in our unconscious minds, an ingrained belief in the magic power of words. By saying that a thing is true we hope to make it true. We shall all become better expositors if we are watchfully aware of this very human weakness.

XIII Qualifications

Qualifications make for a turgid style

Scientists and philosophers, with their conscientious minds, are forever perfecting, modifying, qualifying their first conclusions. They are meticulously careful to avoid over-simplification, to maintain a just balance, to appraise the circumstances so accurately that every one of a number of governing factors is properly valued. This explains why the writing of scientists and philosophers so often contains qualifications.

Unless the author is very skilful the result is a turgid style. For qualifications are among the most difficult things to cope with. The reason is obvious. The work needed to appreciate a scientific or philosophical statement may be considerable. So may be the work needed to appreciate the qualification. The combined work needed to appreciate both at once may well be beyond the capacity of the John Smith whom you are addressing.

The first part of both efforts consists, as I have explained in previous chapters, in the assembly of relevant associations. The metaphorical storekeeper collects these from the shelves of the mind and spreads them out on the floor, as it were. But the associations required for the

Qualifications

qualifications are not quite the same as those required for the statement. So the storekeeper has to collect two sets. If he has to begin to collect the one before he has finished collecting the other, he will become fussed. The floor will be littered with a confused double set of associations. John Smith will be utterly at a loss to know what it is all about. Here is an example:

'Thus foreign nations will be able to transact business with the United States and to accumulate assets there, the above mentioned nations and nationals of other than the excepted nations being, of course, excluded from such transactions. The assets thus accumulated will be free of all but the routine controls and restrictions.'

Never mind what these two sentences mean. We are concerned here with exposition and not with economics. Suffice it that each of the above sentences contains a statement A and a qualification B. In the first sentence the statement A is that foreign nations will be able to transact business with the United States and to accumulate assets there. This short statement is about quite a lot of things: foreign nations, the United States, business transactions, accumulation of assets. Before John Smith can grasp the meaning of the statement he must have collected a substantial number of varied associations. Let it be assumed for the sake of argument that he can do this, and appreciate the significance of the statement, and commit it to memory as fast as he reads it. If he is familiar with the subject the task may not be beyond his powers.

But while he is at work on it the qualification B is introduced. This states that certain nations, described as

The presentation of technical information

'above mentioned' and for which he must cast back his mind to an earlier remark in the article, are excluded. The qualification further mentions nationals of other than the excepted nations. Quite a number of new associations are called for. A renewed effort at understanding has to be made. For the meaning of the qualification is not quite obvious. That is too much to ask of any John Smith; even, I think, if he is a trained economist.

The second sentence might pass. The statement A is that the assets thus accumulated will be free of all controls and restrictions. The qualification B is that they will not be free of *routine* controls and restrictions. If in this second sentence the statement A will be almost obvious to the John Smith whom you are addressing, and it may well be, he will be able to take a reasonably easy qualification in his stride. Suppose he can recall with but little difficulty what the routine controls and restrictions are, then the qualification here is reasonably easy. So for a fairly knowledgeable John Smith the second sentence is well managed. But the first would be bad whoever the John Smith was.

When, in the first draft, you have set down a sentence like that what can you do about it?

Qualification and statement should be clearly separated

One thing is to allow more time to elapse between the statement and the qualification. Get the first well out of the way before you present the second. For there is *some* gain in clarity if you make sure that the storekeeper will

Qualifications

have quite finished his work and John Smith will have grasped the meaning before you trouble him with the new idea that the qualification introduces.

But that alone does not get you out of your difficulties. If you allow the time suggested above, John Smith will not only have grasped the meaning of the statement. He will have also completed the next task and committed it to memory. Then, when the qualification comes along, the storekeeper will have to take the original statement down again from its shelves, unwrap it, and wait for the qualification to be added. Thereupon he must wrap up the parcel anew and put it away once again. You must contrive to save him from such unnecessary and exasperating labour. In other words, you must ensure that John Smith makes a *provisional* mental note of the statement only, and not a *final* one.

There are many ways of doing so. One occurred two paragraphs back, when I used the words 'there is *some* gain in clarity'. The word 'some', helped by italics, hinted that the recommendation to follow immediately would not be a complete one, that more would have to be said before any recommendations could be finally stored away on the shelves. A similar hint is sometimes given in scientific work by introducing a statement with the words 'to a first approximation'.

Alternatively one can introduce the statement with the words: 'One might think A, but there is B.' Thus in the above example: 'One might think that all foreign nations will be able to transact business... But there are exceptions...' Or: 'Some, though not all, nations will be able to... The exceptions are...' Thus managed

The presentation of technical information

both A and B can be expanded as much as may be necessary for clarity without risk of misunderstanding or a premature conclusion being wrongly reached.

Means of avoiding qualifications

However, it is better still to avoid qualifications altogether and this is possible more often that one might suppose. I have said above that qualifications occur frequently in the work of scientists and philosophers because these have frequent occasion to perfect, modify, qualify their first conclusions. What they do can be put into the formula: $A+B=C$ where A is the first conclusion, B the qualification, and C the final conclusion. Whenever such a formula is possible the left-hand side ought not to be presented at all. $A+B$ should be replaced by C. When it is not, the reason is that the author set down the inaccurate statement A *before* it had occurred to him that the statement needed qualifying. Too lazy to rewrite what he had set down already he then added the qualification B, his conscience at rest that he had been 'scientific'. Perhaps he was, but he was also thinking aloud. And that is not a good thing in an expositor.

Here is an example. You are discussing Italian music. You set down the statement: 'All Italian composers have such-and-such a characteristic in common.' Then it occurs to you that this is not strictly true. For it does not apply to Palestrina or any other composers who wrote in the ecclesiastical modes. So you add to your statement A the qualification B: 'Except Palestrina and others who wrote in the ecclesiastical modes.' These composers are

Qualifications

not, I am assuming, relevant to your discourse. You only drag them in because you want to be strictly accurate. If so you should search for a second statement C in which no qualification need be mentioned. After a little thought you may remember that less than a score of years after Palestrina's death in 1594 hardly any significant Italian music was written in the ecclesiastical modes. Your revised draft may then take the form: 'All Italian composers since the beginning of the XVIIth century . . . '

Similar occasions arise frequently in physics or engineering where one can with advantage replace a statement together with its qualification by another statement that does not need to be qualified. Sometimes the reason why it ought to be done is more subtle. As Ohm's Law is one of the laws most familiar to laymen, I will base an example on it. This law states that in a given circuit the current is proportional to the voltage.

Suppose a lecturer says: 'The circuit before you conforms to Ohm's Law, provided allowance is made for temperature changes and the inductance of the circuit.' The first half of the sentence is the statement A. The second half is the qualification B. How should the sentence be altered so as to turn it into an unqualified statement C?

On first thoughts one might argue that such an alteration was impossible and that a qualification here was unavoidable. But second thoughts reveal that, in this instance, the qualification is not only avoidable, it is scientifically incorrect. To make this clear let us suppose that Socrates is a member of the audience.

Socrates: Is it not true that Ohm's Law is a very broad generalisation, valid for a wide range of phenomena?

The presentation of technical information

Lecturer: Perfectly true. It is one of the most universal laws in nature.

Socrates: Then the circuit before us, for which the law only holds with a qualification, is a rare exception, I presume?

Lecturer: By no means. In practice temperature changes and the inductance of the circuit nearly always prevent the current from being strictly proportional to the voltage.

Socrates: Then may I assume that the departure from proportionality is always very minute?

Lecturer: Oh, no. It is often very great.

Socrates: If that is so, circuits must nearly always fail to conform to Ohm's Law. Yet you told me just now that it is one of the most universal laws in nature. A universal law does not have to be qualified in nearly every practical instance.

Lecturer: I did not mean to say that it does. What I meant was that the circuit conforms to Ohm's Law, *together with* the laws that relate current to temperature change and inductance of the circuit.

And the moral of this dialogue? It is that everyone should develop, as a part of his scientific conscience, just such a tiresome little Socrates within himself. His other name is the spirit of self-criticism. When he has become established one should lay before him every qualification that slips from the pen or drops from the lips. He will pass a few. Others, like the example of the Italian composers, he will show to be the result of thinking aloud. Yet others, like the example of Ohm's Law, he will show to be the result of thinking loosely.

XIV Metaphor

Live and dead metaphors

The Concise Oxford Dictionary defines 'metaphor' as the 'application of name or descriptive term to an object to which it is not literally applicable'.

Many words in our language started as metaphors. But the metaphors are now dead and only the words remain. They are like the dead bodies of coral insects that form an atoll in the Pacific Ocean. They serve as raw materials in verbal construction and, as with lumps of chalk, their origin is only recognisable to the trained expert. The dictionary contains such vast numbers of these dead metaphors that one can hardly build any simple sentence without using a number of them. Two or three may go to the formation of a single word, forming respectively a root, a suffix, and a prefix. We owe the most plain and unadorned of our common speech to the fertility of our forefathers in the invention of metaphorical terms of expression.

What began thousands, no, hundreds of thousands of years ago continues today. Contemporary speech and writing still contain live metaphors in large numbers as well as the dead ones. Sitting in a coffee bar one day I overheard a couple of youths discussing some colleague in their office. 'He shot up the ladder with a hell of a lick,' said one. Here metaphors are not so much mixed as crowded. They seem to be built on top of each other as ancient towns, excavated by archaeologists, are found to have been built over each other on the same site, debris from the lower ones reappearing in the construction of those above them.

There is work for scholarship in the study of such colloquialisms. From them one can realise that some of the metaphors now alive are destined, when in due course they will have lost their metaphorical significance, to pass into ordinary plain language with a literal meaning. Thus will our speech ever grow and develop so long as men have imagination enough to clothe their thoughts in metaphorical garments.

This they will have to do, whether they approve of metaphorical language or not, so long as they have any novel thoughts. For it is usually impossible to express what is quite new in any other way. Psychologists, as I have said already, cannot get along without the copious use of metaphor. The ideas with which they are concerned are so new and unfamiliar that no plain words have yet been coined wherewith to express them. Even physicists are sometimes under this necessity, though less often now than when their science was very young. They cannot avoid speaking, for instance, of 'barriers' and 'forbidden transitions' when they do not mean literally that there are barriers or that any prohibitions have been decreed.

Misuse of metaphor

There is no harm in the use of metaphor, even in physics, provided author and reader are both equally clear that the expressions used are to be understood metaphorically and not literally. But confusion of thought and some very naive philosophies will result if it is assumed that, because memories are said to reach consciousness through sometimes more and sometimes less well-worn channels, one may justifiably conclude that memories follow definable passages through the brain; that because one can speak of a storehouse for memories these are separable and could be found in individual brain cells; that any of the spatial metaphors used by psychologists prove thought to have location; that the expression 'laws of physics' implies (now or a long, long while ago at the world's beginning) a legislative authority; that because biochemists have called certain chemical compounds 'organisers' these possess an organising capacity of which any manager might be proud. Such literal interpretations must not be put on metaphors. In choosing the above examples I have purposely avoided those in which the distinction between the metaphorical and the literal meaning of an expression is so obvious that no educated person would confuse the two. I have, instead, selected examples in which educated people, who ought to know better, do very often confuse them.

So when circumstances oblige or encourage you to use a metaphor be quite clear that what you are saying is to be understood metaphorically and not literally.

You may be wise to do this even when you use a metaphor that was once alive but is now dead, such as in physics, law, force, energy, power. For if John Smith is a layman he may not know how very dead the metaphor is. Metaphors do not always lose their power to corrupt thought even in death. They sometimes spring to life in a most disconcerting manner. You tell John Smith that the energy in a kilogram of coal is 25 million joules. Up pops 'energy' and says: 'You remember me, John Smith. We met only yesterday when you said how full of energy your small boy was.' You speak of the law of inertia and 'law' rises up, looking rather saintly and radiant, to remind John Smith of his often expressed opinion that there ought to be a law to make everyone work. The technical terms of physicists are often very evocative. They remind non-physicists of many notions with which they are familiar, of pleasant, interesting notions. But, unfortunately, they are the wrong notions. If layman John Smith only knew how very dead are nearly all the metaphors that physicists use, he would not find popular books on science nearly as exciting as he now does. He would be less inclined to bask in them. But the only losers would, I think, be the writers and publishers of those popular books.

What all this amounts to is that metaphor is, inevitably, a loose inaccurate way of expressing thought. It may convey too much, or too little, or something quite different from what was intended. Should we not then avoid metaphor as tempting us to sloppy thought and careless speech? Should we not, when in an unguarded moment a metaphorical turn of phrase slips from our pens, erase it hastily and labour to find the proper plain words into which to translate it? Even though we should not then do for posterity what our more imaginative and (is it thought?) less seriously minded fathers did for us—enrich the English tongue; even though we should not serve the future as well; should we not serve the present better by ruthless banishment of that mischievous treacherous will-o'-the-wisp, metaphor?

The value of metaphor

Certainly not. True, metaphor has its dangers and will be avoided by the very cautious. True, some of the worst writing is crowded with metaphor. But so is some of the very finest. If those who are too sparing with its use will never reach the extremes of commonplace and vulgarity to which the misuse of metaphor may carry an author, they do, nevertheless, reveal mediocre barren minds. They are dull writers and not at all easy to follow.

And this holds not only for imaginative literature, but also for functional writing; for besides providing the most attractive and stimulating way of saying things, metaphor also sometimes provides the clearest and most concise way. Even after hundreds of thousands of years of development, language is still but a clumsy instrument for so fine a thing as the thought of man.

It would rarely make for clarity if the purpose of the discourse were only to *express* information. But the purpose, be it remembered, is also to *convey* the information from mind to mind. This can often be done by a word with the help of a metaphor where a hundred words would be required without its help. And the hundred

The presentation of technical information

plain words would confuse, while the one metaphorical one clarifies. So the justification for the use of metaphor in functional writing depends on how much more, in addition to expressing the information, need be done in order to convey it.

If the information is conveyed as readily as it is expressed, nothing more is needed; metaphor can serve no useful purpose and it would be out of place. Consequently this device will hardly ever be found necessary when the information is purely factual. There are no live metaphors in a railway timetable or the telephone directory. Nor are there any in a legal document. There ought not to be many in a report that describes a scientific experiment. But there may be quite a large number in a discourse in which arguments are weighed against each other, a reasoning process is placed on record, implications are discussed, or new and unfamiliar ideas are introduced.

An article in a weekly periodical happens to lie open before me at the moment; crowded together occur the following: 'a cloak of unity, however threadbare', 'bargaining counter', 'water down the plan', 'dug in their toes'. 'Formula' is, moreover used in this article to mean a carefully prepared statement of policy to be submitted for agreement. Elsewhere I see the words 'ceiling' and 'target'. If one were to translate these expressions into dead metaphors one would have to say respectively 'the highest attainable limit' and 'the figure that has been laid down as the one to strive for'.

The other metaphors just quoted could, of course, also be translated into plain English. The result would be that

Metaphor

the article in which they occur would be longer, more cumbersome, less clear, less memorable; in a word, less functional.

Characteristics of a good metaphor

Thus metaphor should be used only with great care and discretion; but it should be used, even in functional writing. When a discourse is spoiled by its metaphors it may not be because there are too many of them. It may be because those that are there are bad ones. And the search for good ones is not always easy. It is worth while to take considerable trouble with it and to reject many bright ideas before retaining the best. This best possesses, I think, three desiderata: aptness, freshness, and familiarity.

The most apt metaphor is the one that remains apt when expanded. Thus in an example quoted above 'cloak of unity' is expanded by the further notion 'however threadbare'. The expansion adds greatly to the quality of the metaphor. It makes it more attractive, more stimulating, more illuminating. What a lot of work a really apt metaphor does. It defines the context, conveys a message, makes it acceptable, makes it comprehensible, helps the work of memorising, all in one or two words.

A metaphor, however apt, can never do so much if it lacks freshness. And some metaphors are far from fresh. They might be described as moribund. They have been worked so hard that they are nearly worn out. The 'avenues' that politicians explore, the 'stones' that they do not leave unturned and their 'blueprints' have long reached this stage of decrepitude. So has the 'angle' from which a proposition is viewed. Perhaps this one is dead already. For it is hardly metaphorical to say 'These facts constitute my angle. What is yours?' as I have recently heard. I hope that, when they are at long last quite dead, avenues, blueprints, stones, and angles may never become literary building materials.

Other metaphors seem to wear better. 'Target', 'ceiling', the various 'levels' at which political decisions are nowadays taken, 'bull's eye', 'bargaining counter', 'trump card', 'red herring' have had a 'long run'. Yet they have enough 'vitality' still to appear 'in the prime of life'. Perhaps it is because they remain apt even when greatly expanded. When they are ready to pass on into the next world, the world of plain language, they may perhaps do so without damage to the English tongue. It has suffered worse incursions without taking harm.

The freshest metaphors are those that spring newborn from an author's brain. But they are also the most treacherous. For they do not always meet the third desideratum, familiarity. Some of them are like family jokes, comprehensible to the few and quite meaningless for the many. If this is so, neither their aptness nor their freshness can make them functional. They may be justifiable in the work of poets whose merit is their obscurity. I do not know. But to be functional they must be culled from a field in which our John Smith is wont to wander. A specialist addressing fellow specialists may, with excellent effect, find some of his metaphors in their common field of study. That such metaphors are like family jokes only adds to their usefulness. It gives to all

Metaphor

the John Smiths who are being addressed a sense of unity among themselves and with the author; almost a sense of exclusiveness, if not of superiority, over those outside the circle of specialists. But when addressing others who are not members of the specialist circle, an author will be wise to go to fields that the common folk frequent. The best metaphors are usually derived from the most familiar sources.

And, moreover, a good author does not invent *all* his metaphors. Universality is often better than freshness. So most of those that he employs, though not too hackneyed, are already well known. They belong to our common language. Well placed and well spaced, a few brand-new metaphors are delightful. Many crowded together become very exhausting.

When a metaphor is at once fresh and apt and easily recognisable by everybody, and when it is also in a conspicuous setting where one has time and opportunity to appreciate it, what splendid economy in words it achieves! A master of language like Winston Churchill can, with its help, in one revealing flash illuminate what any number of plain words would leave in a dull shade. Discussing the Italians' claim for recognition by the Allies just after their defeat he says: 'They must first work their passage'. In half a dozen words he defines a situation, expresses a judgment on it, and convinces his hearers that the judgment is sound. And all this in such a form that it fits instantly into a permanent place on the shelves of the memory.

A subtle touch is, moreover, that, addressing a seafaring nation, he uses a seafaring metaphor. To the

The presentation of technical information

British people 'work their passage' has just enough private significance, just enough exclusiveness, to convey to them that sense of being united as a family, with the loyalties and responsibilities that unity calls for. Churchill selected here a metaphor that can accomplish on a sublime plane what is done on a trivial one by the family joke.

XV Words

Proper use of technical terms

Humpty Dumpty made words mean whatever he wanted them to. But Humpty Dumpty had a great fall.

Engineers have no need to follow his example. For the British Standards Institution has for years been responsible for the technical vocabulary of the profession. Constantly watchful, it has been able to establish suitable technical terms before unsuitable ones had gained currency. The result is a living, growing, flexible, unambiguous vocabulary, in which every word can be justified scientifically, logically, and linguistically.

In such a vocabulary each word has only one meaning and each meaning only one word; this meaning is the same for everyone who uses the vocabulary; and this meaning is available to everyone as a definition in a published glossary. These glossaries should be employed rigidly in all technical work. They provide engineers with a most useful tool.

They also impose a most salutary discipline. For, though technical terms form only a small fraction of the words used in any piece of writing, the precision with

The presentation of technical information

which they have to be employed has, I think, had a good effect on the thinking and writing of engineers. When he is forced to exercise extreme care with some of his most important words an author will acquire the habit of exercising similar care with others.

Those engaged in imaginative literature are not subject to this particular discipline. They use their words with the same care as a conscientious technologist does. But their principle of selection is different. It may be euphony, allocation of emphasis, creation of atmosphere, definition of mood. It can only be applied successfully when more than one word is available for the same thing. So imaginative writers have developed a language rich in synonyms. With the help of such a language an author can accurately define the just perceptible difference in shade between two meanings.

It is understandable that to those who dispose of such a profusion of words a language from which synonyms have been ruthlessly excluded must appear barren, colourless, mechanical. These epithets may be just; but such a language is exactly what the scientist and technologist require.

If language serves to *express* thought, it also helps sometimes to *create* thought, as I have said already. We often have to speak before we can find out what we think. The habit of our thinking is fashioned by the kind of meaning we habitually attach to words. If we use words habitually that serve to express finely shaded grades of meaning our thought will become subtle, but may come to lack logic. If we use words habitually that serve to express shadeless, unambiguous meanings our

Words

thought will become logical, but may come to lack subtlety.

So the complete man should get some practice with both languages. But in these pages it is more important to emphasise the value of a language that lacks synonyms than that of one rich in them. This is all the more important because those who do not properly appreciate the value of the former are not all poets (for whom there is every justification), but also too often scientists and philosophers.

The wrong employment of the words 'force', 'power' and 'energy' provides an example. To an imaginative writer, no doubt, these words are roughly interchangeable, though he would expect to meet occasions when one of them would express a given shade of meaning more accurately or more neatly than either of the other two. But a physicist and an engineer know that there can be no such occasions. For the words are not synonyms. And the meanings they express cannot be shaded. When the word 'force' is the right one it is also the only available one. It would then be just as wrong to say 'power' or 'energy' as to say 'jabberwock'.

Those who use these and other technical terms wrongly are, of course, ignorant of their correct meaning. But I suspect that the defect is deeper than mere ignorance.

They use these words as though they were synonyms because they do not know that a word can ever have that hard, clear, unambiguous, unique meaning that is given to words in a technical vocabulary. They do not even know that thought can ever be so hard and clear and unambiguous.

The presentation of technical information

Proper use of words in general

So much for technical terms. As they form only a small fraction of the words used in any discourse a glossary is not sufficient to ensure that the right word shall be used on every occasion. It can only set the standard. This, as I have suggested above, it may have done to some small extent in the engineering world. But there remains both in that world and elsewhere much to be done. Persons who have something to say, from Fellows of the Royal Society downwards, often find the task of putting it into words an irksome one. So they tend to grow careless. And if the great ones lend the authority of their distinguished names to carelessness the lesser ones will feel secure from censure if they are careless too. If indifference to the right choice of word continues for long enough we shall eventually decline to the stage where words are used so loosely that no one will know what anyone else is trying to say. At present we have only sunk to the level where we think we know and are usually right, but are not always quite sure.

The examples in the list to follow are limited to those in which the objectionable word conveys the wrong meaning. It does not contain examples in which the word conveys the author's intention in an ugly or a clumsy way. Some of the examples seem to result from mere carelessness, some from sheer illiteracy, some from inability to think logically, and some from ignorance of technical matters.

If they often appear to be rather extreme it is because of the prominence they receive when attention is drawn to them in a list of errors. They are, nevertheless, typical. Were we to meet them in their context, in a learned book or a newspaper article, say, we should pass over three-quarters of them, at least, without noticing that the wrong word had been used, so insensitive have we become to the misuse of words. If such insensitiveness is allowed to persist we shall continue in the present vicious spiral where low standards encourage further carelessness and this, in turn, further depresses the standards. To get out of it we must train ourselves to become more observant of the unfunctional use of words, whether it occurs in other people's work or our own. Lists like the one given below may show the sort of error one should look out for.

Carelessness

'The coal is stored in two ways.' It is not stored in ways. Possibly it is stored in two places. Or possibly two distinct storage methods are in use.

'The blowdown consists of periodically draining off some water from the boilers.' Blowdown is a name for the water drained off, not for the process of draining.

'The water has been purified in the screens.' The water is not pure after it has passed the screens. Screens cannot purify water. They can only remove large solid objects.

'The evaporator is to provide a supply of distilled water.' This in order to describe the purpose of the evaporator. It is all too common to expect 'is' to do more work than even Humpty Dumpty could have got out of it. Here it stands for 'has been installed'.

In the following example it is the other way about.

The presentation of technical information

'The efficiency depends on the ratio of the amount of energy generated to the potential energy in the oil.' The efficiency does not *depend* on this ratio; it *is* this ratio.

'Had' is sometimes worked as hard as 'is'. 'The condensers had a vacuum of 750 mm of mercury.' 'Had' should be replaced by 'were working at'.

Illiteracy

'The importance of this item has been diminished by introduction of the grid.' 'Diminished' should be 'reduced'.

'Does the responsibility for the war rest with the politicians?' Meaning: 'Are the politicians to blame?'

'It would be a grave responsibility to conclude a separate peace.' Meaning: 'a grave mistake'. Misuse of the word 'responsibility' has become almost a fashion.

'This is calculated to affect the opposition more than the government.' 'Calculated' is nowadays being used instead of 'likely' with distressing frequency. Moreover 'affect' is here apparently taken to mean 'affect unfavourably'.

'The Battersea Power Station has achieved two of these results.' Should be: 'In the Battersea Power Station two of these results have been achieved.'

'Two of the machines are of special design. The others are of the standard type.' It should probably read 'are of the conventional type'. Failure to distinguish between the meanings of 'standard' and 'conventional' is not uncommon.

'There are many advantages to be looked for when

choosing the site for a factory.' 'Advantages' is often used in this wrong sense.

'In my opinion the former method is preferred and should be adopted.' 'Preferable' is meant.

'We hastened to our next object.' Meaning: 'objective'.

'What would be the outcome if any part of the factory installation were to break down?' A surprisingly large number of people have come to believe that 'outcome' means the same as 'result' or 'consequence'.

Illogical thinking

'The problems of ash disposal are dust nuisance, corrosion, and poisonous gases.' These are not problems, though they create problems.

'A window in the control room enables the operator to see the proper functioning of the plant.' It is to be hoped that he would also be able to see when the plant was not functioning properly.

Ignorance of technical matters

Many of the examples given above have been plucked from the field of engineering. Lest it be thought that engineers alone need advice in the proper use of words I will quote one from another field. It is the most fertile of all for examples of unfunctional writing and lies on the border between science and philosophy, where, I am sorry to say, any old thing is deemed good enough. It combines inability to think logically with a surprising lack of knowledge of physics.

'So, similarly, under physical analysis, item after item

The presentation of technical information

of the behaviour of the whole is resoluble into behaviour of the subvisible physical units. These latter our fancy can, perhaps, think of ultimately as electrically charged "packets of motion".' The passage purports to enlighten the reader on the general nature of any physical system. I suggest that none of the following words can bear investigation: resoluble, behaviour, physical analysis, ultimately, packets, fancy, perhaps. Let them be considered in that order.

Resoluble. Can one resolve behaviour? One can study it, describe it, explain it. But it is meaningless to say that one resolves it. 'Can be described' seems best to convey the author's meaning.

Behaviour. I suggest that the proper word is movement. Behaviour is not possible without movement, admittedly. But it is not the same thing. Behaviour refers to a specific or characteristic kind of movement, by which it can be distinguished from other kinds. One speaks of the behaviour of a cat or a child. One does not speak of the behaviour of the wind or the waves. There is nothing to show that for describing what happens to the subvisible particles of matter the word one applies to a cat or a child is better than the one applies to the wind or the waves.

Physical analysis. Is whatever is said in this sentence true only under physical analysis? Whatever the system may be to which the author is referring, the movement (or behaviour) of the whole can be described as (or is resoluble into) the movement (or behaviour) of its constituent physical units. I cannot see how the statement would be less true, or less informative, if the words 'under physical analysis' were omitted.

Words

Ultimately. I can attach no meaning whatever to this word in this place.

Packets. Whether the word be meant literally or metaphorically packet implies items in some sort of container. One can have a packet of peas or a packet of books. Metaphorically one can have a packet of troubles. But an abstract generalisation like motion cannot be contained in a packet. If the statement is to be understood literally it must be wrong. If metaphorically, the metaphor is a peculiarly bad one.

Fancy. Our fancy can do a great deal. It can think of the subvisible physical units as pink elephants. But one thing our fancy cannot do is to conceive an impossibility like a packet of motion, electrically charged or otherwise. Apart from this it is not in the least informative to say how our fancy can *think* about things. The important point is what the things *are*. This we are not told.

Perhaps. This word in this place seems to convey nothing.

When the right words have been substituted for the wrong ones and the passage has been suitably remodelled it amounts to this: 'In any system the movement of the whole is the sum of the movements of the component parts. The smallest of these are always subvisible physical units of which it is only known that they carry electric charges and are in motion.' When thus rendered comprehensible the passage does not sound very profound. Indeed a reputation for profundity is often built on nothing more substantial than a grossly unfunctional style.

XVI Presenting numerical information

The first fifteen chapters of this book have been concerned with the psychology, grammar, and words that can be used to advantage in presenting non-numerical technical information. There is an equally important grammar and style in the presentation of numerical information. This is particularly important now that the scientific and technical communities throughout the world have adopted a new and consistent system of units.

Throughout its history, science has had to rely on a hotchpotch of units, and systems of units, to express the magnitudes of physical quantities. Traditionally, for example, the Imperial foot-pound-second system has been used in Britain and America. On the continent, the metric centimetre-gram-second system has been in use since its introduction under Napoleon. Both these systems have now been abandoned and replaced by SI units (Système International d'Unités). The decision to adopt this system was taken as long ago as 1960 by thirty nations meeting at the Conférence Générale des Poids et Mesures. In 1968 the Royal Society recommended that SI units should be used exclusively in scientific and technical journals, and that they should be introduced in

Presenting numerical information

school curricula. At last, therefore, scientists and technologists throughout the world can talk to each other in a quantitative language that they can all understand without the need for conversion factors, slide rules, and a great deal of totally unnecessary irritation. There can now be no doubt that numerical information, in a technical context, should always be presented in SI units and that the rules laid down for the use of SI units should always be followed.

SI units

This system is extremely simple. It is derived from the metre-kilogram-second system and has the following seven base units, which are the only units to have formal definitions. The last two units given below, the radian and the steradian, are dimensionless and are called supplementary units.

Physical quantity	SI unit	Symbol for unit
length	metre	m
mass	kilogram	kg
time	second	s
electric current	ampere	A
thermodynamic temperature	kelvin	K
amount of substance	mole	mol
luminous intensity	candela	cd
plane angle	radian	rad
solid angle	steradian	sr

The presentation of technical information

Units for all other physical quantities are derived from these nine units. The system is coherent, i.e. all the derived units are obtained from the base units without the use of numerical factors. For example, the base units of length and time are the metre and second respectively. Therefore the derived unit of velocity is automatically metres per second. This unit has no special name. However, the derived unit of force, on the basis of Newton's second law of motion, is the kg m s^{-2}. This unit has the special name *newton*.

The derived units having special names are given below.

Physical quantity	SI unit	Symbol	Equivalent in base units
frequency	hertz	Hz	s^{-1}
energy	joule	J	kg m^2 s^{-2}
force	newton	N	kg m s^{-2}
power	watt	W	kg m^2 s^{-3}
pressure	pascal	Pa	kg m^{-1} s^{-2}
electric charge	coulomb	C	A s
electric p.d.	volt	V	kg m^2 s^{-3} A^{-1}
electric resistance	ohm	Ω	V A^{-1}
electric conductance	siemens	S	Ω^{-1}
electric capacitance	farad	F	A s V^{-1}
magnetic flux	weber	Wb	V s
magnetic flux density	tesla	T	V s m^{-2}
inductance	henry	H	V A^{-1} s
luminous flux	lumen	Lm	cd sr^{-1}
illumination	lux	Lx	cd sr^{-1} m^{-2}

Presenting numerical information

In SI units there is only one base or derived unit for each physical quantity. In older systems there are often a bewildering profusion of units. For example, in the foot-pound-second system lengths may be expressed in inches, feet, yards, poles, chains, furlongs, or miles (to name only the most common). In SI units only the metre or its sub-multiples or multiples are used. Thus short lengths are expressed in *milli*metres, long ones in *kilo*metres. The following prefixes are used to indicate these decimal sub-multiples or multiples with all base or derived SI units. No others and no combinations of them should be used.

Factor	Prefix	Symbol	Factor	Prefix	Symbol
10	deca-	da	10^{-1}	deci-	d
10^2	hecto-	h	10^{-2}	centi-	c
10^3	kilo-	k	10^{-3}	milli-	m
10^6	mega-	M	10^{-6}	micro-	μ
10^9	giga-	G	10^{-9}	nano-	n
10^{12}	tera-	T	10^{-12}	pico-	p
			10^{-15}	femto-	f
			10^{-18}	atto-	a

In general, prefixes involving powers of three are preferred, e.g. 10^{-3} milli-, 10^3 kilo-, 10^{-6} micro, 10^6 mega-, etc.

Conventions adopted with SI units

When SI units were adopted a number of conventions were recommended to facilitate their use. These rules

The presentation of technical information

should be obeyed in the presentation of technical information, in the same way as the rules of grammar and spelling are obeyed. They constitute a numerate style, just as grammar and spelling constitute a literate style.

The first rule is that the magnitude of a physical quantity must always be treated as the product of a pure number and a unit: i.e. physical quantity = number × unit.

For example, the physical quantity *velocity* (symbol v) may be expressed as, say:

$v = 50$ m s^{-1}

To write $v = 50$ is an incomplete statement. It would be equally wrong to refer to v m s^{-1}; by substitution in the above equation this would mean, in this case, 50 m s^{-1} m s^{-1}, which is not intended.

This point is particularly important in labelling the axes of graphs and heading the colums of tables. A curve on a graph relates numbers only: for example, if a potential difference (V) measured in volts is to be plotted against a current (I) in milliamps, the axes should be labelled V/V and I/mA. This complies with the equations:

$V = n$ volts   and $I = m$ milliamps
or V/volts $= n$ and I/milliamps $= m$
where n and m are pure numbers.

If the level of the audience or readers is rather more elementary than this, the axes or columns may be correctly labelled:

V in volts   and I in milliamps.

In the printing or writing of SI units certain conventions are adopted:

Presenting numerical information

(1) Units may be written out in full or the agreed symbols may be used; no other abbreviations should ever be used and the letter s is never added to symbols to indicate a plural. Symbols for units are always written or printed in roman type, whereas symbols for physical quantities are always printed in italic type or written with a line underneath them to indicate italics.

(2) A full stop is not used after a symbol for a unit, unless it occurs at the end of a sentence.

(3) Names of units, even when they commemorate a person, never have an initial capital letter when written out in full, e.g. newton, hertz, metre, etc. However, the symbol for a unit, the name of which commemorates a person is either a capital letter or has an initial capital letter, e.g. N for newton, Hz for hertz, etc. All non-personal names of units have lower case symbols.

(4) When two or more symbols are combined to indicate a derived unit, a space is left between them. No space is left between a prefix indicating a sub-multiple or multiple and the symbol for the unit. For example, m s indicates metre seconds, whereas ms indicates milliseconds.

(5) When derived units include a quotient, e.g. metres per second, they are best written in the form $m\ s^{-1}$. They may also be written using a solidus, e.g. m/s, but a solidus must never be used twice in the same unit. Thus, acceleration in SI units has the units $m\ s^{-2}$ which may be written m/s^2, but never m/s/s.

The presentation of technical information

(6) As already stated, symbols for physical quantities are printed in italic type to distinguish them from the roman type of the symbols for units. For example, *V* stands for potential difference and V for volts. Symbols for vector quantities are printed in bold face italic type when it is desired to draw attention to their vector nature (e.g. ***F*** for force). Symbols for tensors of the second order should be printed in bold face sans serif italic type (e.g. ***S***).

(7) Two decimal signs are now internationally accepted, the on-line full stop, e.g. 5.6, and the on-line comma, e.g. 5,6. The former is recommended in the UK and US; the raised point, e.g. 5·6, should be discontinued.

In order to avoid confusion between the comma used as a decimal sign and the comma used to divide large numbers into groups of three digits, the latter practice is no longer in use. In place of the comma a space is left, e.g. 563 240. This practice is only recommended when there are more than four digits, in the case of four digits no space is left, i.e. 5632 not 5 632. Whenever possible the standard form should be used with large numbers, e.g. $5.632\ 407 \times 10^6$ instead of 5 632 407.

If these simple rules are followed a great deal of confusion can be avoided.